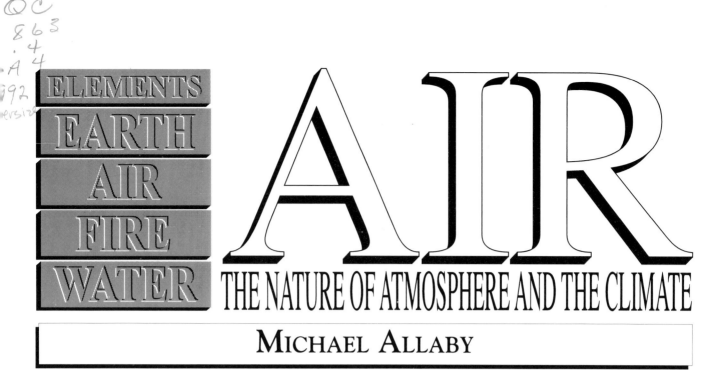

ELEMENTS
EARTH
AIR
FIRE
WATER

AIR
THE NATURE OF ATMOSPHERE AND THE CLIMATE

MICHAEL ALLABY

Facts On File
New York • Oxford

ELEMENTS: AIR
Copyright © 1992 by Michael Allaby

All rights reserved. No part of this book may be reproduced or utilized in any
form or by any means, electronic or mechanical, including photocopying,
recording, or by any information storage or retrieval systems, without
permission in writing from the publisher.
For information contact:

Facts On File Limited   or   Facts On File, Inc.
Collins Street   460 Park Avenue South
Oxford OX4 1XJ   New York NY 10016
UK   USA

A British CIP catalogue record for this book is available from the British
Library.

A United States of America CIP catalogue record for this book is available
from the Library of Congress.

ISBN 0-8160-2525-8

Facts On File books are available at special discounts when purchased in bulk
quantities for businesses, associations, institutions or sales promotions. Please
contact the Special Sales Department of our Oxford office on 0865 728399
or our New York office on 212/683-2244 (dial 800/322-8755 except in NY,
AK, or HI).

Produced by Curtis Garratt Limited,
The Old Vicarage, Horton cum Studley, Oxford OX9 1BT

Technical artwork by Taurus Graphics Limited
Pencil drawings by Paula Chasty

Printed in Hong Kong

10 9 8 7 6 5 4 3 2 1

This book is printed on acid-free paper.

(Previous pages, left) The Space Shuttle *Discovery* is launched
from the Kennedy Space Center. Among the tasks carried out
by space shuttles on their various missions is photographing
Earth — in this case, using a hand-held camera aimed through
the 'ceiling' windows of the craft. (Previous pages, right) Here,
large cumulonimbus clouds can be seen building up over Zaire.
Because of the haze created by agricultural burning, the ground
cannot be seen.

CONTENTS

INTRODUCTION

Water, air, earth and fire have been known throughout most of our history as the 'elements'. No one knows when these elements were first defined, but they were certainly being discussed around 600 BC and it was such discussions, of ideas that were developed and elaborated over centuries, which laid the foundations of the particular way of examining and thinking about natural phenomena we now call 'science'.

The word 'element' is used differently today, but its ancient meaning continues to haunt our language and culture. It suggests balance and justice, concepts which seem appropriate to our modern concern about the condition of the natural environment.

This is one of a series of four books, each of which describes one of the ancient elements, but in a modern way. You will read accounts of the substances which compose each of the elements, how they are formed, how humans have made use of them, ways in which we may be damaging them, and the steps that are being taken to render our exploitation of them more rational. The text deals with scientific and technological concepts, but places them in a cultural and historical context that to some degree restores the older images their names still conjure, even now, as we approach the end of the twentieth century.

In the sixth century bc, in Miletus, a flourishing, cosmopolitan, commercial town in Asia Minor, a number of Greek philosophers, comprising the Milesian school, developed the ideas that formed the basis of the ancient concept of the elements. Thales, who predicted an eclipse that is known to have occurred in 585 BC, maintained that water is the original substance from which all others are derived. Anaximander (about 610-547 BC) disagreed. He held that the primal substance, from mixtures of which all others are composed, is eternal and indefinable, and is transformed into the substances we see around us. Each of the elements — earth, fire, air and water — is a god, seeking to expand his dominion, but this proves impossible because the proportion of each element is fixed and the balance cannot be disturbed. Fire produces ash, which is earth. When warmed, water becomes air and, when cooled, air turns to water.

Anaximenes, a younger contemporary of Anaximander, maintained that the primary substance is air. The soul is air, he said, fire is a rarefied form of air and, if air is condensed, it turns first into water, then earth and finally into stone. These arguments over the primacy of the elements continued until Empedocles, who lived around 440 BC, resolved the matter by proposing that the four elements were of equal importance.

Some three centuries later, the philosopher Aristotle (384-322 BC) developed the idea further. He maintained that the elements are not eternal, but change from one to another and mingle under the influence of the approach and retreat of the Sun, and everything is in constant motion. Fire is absolutely light and moves upwards, air is relatively light, earth is absolutely heavy and moves downwards, and water is relatively heavy. All matter which exists below the Moon is composed of the four elements, and subject to change and decay, while matter above the Moon is made from a fifth, eternal element.

The ideas of Aristotle, expounded in his many writings, dominated European thought for 2000 years, and their influence was not wholly benign. Indeed, they became such a serious constraint on intellectual development that eventually innovative philosophers and scientists found themselves beginning their expositions by refuting one or another Aristotelian doctrine. Yet it was Aristotle who insisted that the route to understanding natural phenomena begins with detailed observation of them, which is the principle underlying all modern science.

Gradually, old ideas gave way to new. William of Occam (*c*.1290-1350), though an Aristotelian, contributed the maxim, known as 'Occam's razor', that explanations should be free from unnecessary hypotheses — the simplest explanation is likely to be the best one. Copernicus (1473-1543) and Galileo (1564-1642) demolished the Earth-centred view of the universe, and Johannes Kepler (1571-1630) discovered the laws of planetary motion. Francis Bacon (1561-1626) began the process of developing an

(Opposite) Aristotle
(**384-322** BC)

inductive scientific system by which general laws may be derived from observation and experiment. It was Robert Boyle (1627-91), regarded as one of the founders of modern chemistry, who, in 1661, first used the old word 'element' to describe a substance that cannot be separated into two or more simpler substances.

These books, then, aim to bridge the historical gap between ancient and modern views of the world. In *Air*, the second title in the series, the story is told in three parts. The book begins by describing the composition of air, where it comes from, and the evolution of the atmosphere, contrasting the atmosphere of our planet with those of Mars and Venus. It goes on to describe the layered structure of the atmosphere, the reason it is layered, and the conditions that are found in each of the layers.

The second section is concerned with the movement of air, the effect on it of the rotation of the planet and the global circulation of the atmosphere. This leads into a general description of climates, their present distribution and their history, with particular reference to the ice ages.

The final section of the book begins with a description of the radiation Earth receives from the Sun, the way this is influenced by the atmosphere and the causes and consequences of air pollution. This is followed by an outline of the evolution of flight and the history of human aviation, and the book ends by describing the way the movement of air has been exploited as a source of energy.

(Right) Galileo
(1564-1642).
(Opposite) Francis Bacon
(1561-1626).

WHAT IS AIR MADE OF?

In the early summer of 1941 the first British jet-powered aircraft, a Gloster E28/39 airframe fitted with a Whittle W-1 engine, underwent its test flights and, according to the legend that became popular among aviators, it was seen by several fighter pilots. When they returned to their base, the pilots described a curious machine that clearly flew, but that equally clearly lacked the propeller without which its flight seemed to them impossible. As they debated the apparent paradox, several solutions were proposed. One was that the engine was a kind of pump, which worked like a vacuum cleaner, and sucked the aircraft through the sky.

This attempt at an explanation was regarded as a joke by fliers who, just a few years later, were very familiar with jet engines. There is a timeless quality in such mockery of the ignorant by the informed, but the real joke is more profound, for the absurd 'explanation' is based on the very ancient fallacy that 'nature abhors a vacuum'. We repeat the fallacy whenever we use the term 'vacuum cleaner', for the cleaner does not create a vacuum, only a flow of air, and the dirt it collects is blown, not sucked, into the bag, just as a jet aircraft is pushed, not pulled, by its engine.

The idea that a vacuum cannot exist in nature follows directly from the view that all real objects are made from one of the four elements — earth, water, air and fire (beyond which lies a fifth substance, 'ether') — either in its pure form or from some mixture of two or more of them. A vacuum is empty, consists of nothing at all, and so it cannot be real. This was a point on which Aristotle was most insistent, and it is mainly from him that the popular idea about the vacuum is inherited. It was not until the seventeenth century that the fallacy was demonstrated by the Italian scientist, Galileo (1564-1642), who also showed that air has weight and that it offers resistance to bodies moving through it. He maintained that it was because of this resistance that a feather falls more slowly than a stone, although the force of gravity acts equally on both.

The weight of air

Galileo showed that air has weight by weighing it. He pumped into a container two or three times its own volume of air, and weighed the container. Then he released the pressure, allowed the compressed air to escape, and weighed the container again. The difference between the two weights represented the weight of the air he had added. He made a vacuum by drawing an airtight stopper up a glass cylinder containing water. He argued that the space between the end of the stopper and the surface of the water had to be a vacuum, because it was impossible for it to contain anything. He was mistaken — even assuming no air could leak past the stopper, the space would have contained water vapour — but he made his point.

People had known for centuries — the Romans studied the phenomenon — that it is impossible for a suction pump to raise a column of water more than 34 feet (10.4 m), but no one knew why and experiments were hampered by the difficulty of constructing such long tubes. The vacuum experiment renewed interest in the subject, and Galileo had proposed a way of simplifying the work by suggesting that the maximum height of the column was inversely proportional to the specific gravity of the

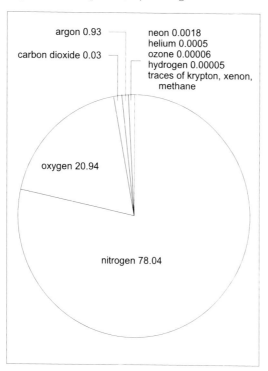

argon 0.93
carbon dioxide 0.03

neon 0.0018
helium 0.0005
ozone 0.00006
hydrogen 0.00005
traces of krypton, xenon, methane

oxygen 20.94

nitrogen 78.04

Composition of the lower atmosphere by percentage volume

liquid used. This means that a heavy liquid can be raised a much shorter distance than a light one, such as water, and so the experimenter can use a much shorter tube.

In 1643, Evangelista Torricelli (1608-47), an assistant to Galileo, tried mercury. He filled a narrow cylinder with mercury, sealed the open end with his finger, inverted the cylinder into a bath of mercury, removed his finger, and the mercury level in the cylinder fell, leaving a vacuum above it. The height of the mercury column will have been about 30 inches (762 mm), because Torricelli had invented the barometer.

Because of the way he conducted his experiment, the mercury did not rise up the tube, it fell back down a tube he had filled to the brim. He realized, and was the first to do so, that the mercury is forced up the column not by suction from above, but by the weight of the air pressing down on the liquid in the bath — a pressure to which the mercury in the cylinder is not exposed, because it is sealed from the air outside. If air resists the movement of objects through it, and if it weighs heavily enough to force mercury to rise up a tube, it must be much

more substantial than earlier thinkers had supposed.

Blaise Pascal (1623-62), the French scientist and mathematician who turned, later in his life, to writing on religious subjects, set out to test the theories of Galileo and Torricelli. Pascal suggested that the atmosphere might be likened to an ocean. If that is so, the weight of air between an observer and the top of the atmosphere should be less at a high altitude than at sea level, because the overlying layer of air is thinner, and Torricelli's barometer might measure the difference. Pascal made many barometers and found that the height of the mercury was always the same. The next step was to measure the weight of air — the air pressure — at height. He could not do this himself, for Pascal was a semi-invalid, so, in 1648, he asked his brother-in-law, Florin Périer, to perform the measurement for him, on the Puy de Dôme, a mountain in central France that is 4806 feet (1465 m) high. Périer found that a barometer set at the foot of the mountain showed the same reading all day, but measurements taken on the mountain recorded a pressure

Mudflats provide habitats for the bacterial populations that release gases involved in the cycling of elements between water and land. These flats are in Norfolk, England.

that decreased with increasing height. Pascal's theory was confirmed, and it proved that air is a real material substance.

When scientists find the answer to a question that had puzzled them, it often happens that the answer raises several new questions to take the place of the one they have answered. This was no exception. Establishing that air is a substance immediately led scientists to ask what kind of substance it might be. Is air pure, for example, so it cannot be broken down into any simpler substances, or is it a mixture of substances, each of which can be identified separately?

Robert Boyle (1627-91), the English natural philosopher, and his assistant, the English physicist, Robert Hooke (1635-1703), discovered other properties of air. Their work established the relationship between the pressure and volume of a gas when the temperature remains constant — which became known as Boyle's law — and, using pumps made by Hooke, Boyle discovered that air, or some constituent of air, is essential to living animals. He found that an animal will die if it is kept in a sealed chamber, but if the chamber is opened to allow air to enter before the animal expires, it will recover fully. He also found that if a burning candle is sealed inside an airtight jar, after a time the flame will be extinguished. Boyle had almost discovered oxygen, although he was unaware of it.

Composition of air

The first constituent of air to be identified was 'gas sylvestre', which we know as carbon dioxide. This gas was discovered by a Belgian chemist and physician, Jan Baptista van Helmont (1577-1644). He had discovered the existence of gases distinct from the air itself, and claimed to have coined the word 'gas'. Gas sylvestre, he found, was given off by fermenting liquor and also by burning charcoal, but it was a Scottish chemist and physicist, Joseph Black (1728-99), who described, in his doctoral thesis in 1754, that gas sylvestre is given off when magnesium carbonate (known to Black as magnesia alba) is heated, and that this gas is distributed in the atmosphere. Because it can be held in a variety of solids, he called it 'fixed air'.

The carbon dioxide (gas sylvestre) molecule consists of one atom of carbon (C) bonded to two atoms of oxygen (O). The chemical symbol for the compound is CO_2, and the gas accounts for about 0.03 per cent of the lower atmosphere, by volume, although the proportion is increasing, mainly because of our combustion of carbon-based fuels (see page 157). The carbon dioxide is produced by combining carbon and oxygen, the oxygen is derived from the air, and so the increase in the proportion of carbon dioxide is accompanied by a similar decrease in the amount of oxygen — the total amount of gas in the atmosphere is not increasing.

Oxygen was first discovered, in about 1772, by the Swedish chemist Carl Wilhelm Scheele (1742-86), who had also discovered chlorine (in 1770). Scheele described 10 ways to produce oxygen gas (he called it 'fire air'), and showed it to be essential to aquatic organisms, as well as those which live in air. He wrote one book, *Air and Fire*, in which he maintained that air is composed of two gases, one of which inhibits combustion ('foul air') and the other, 'fire air', which supports it, but he waited until 1777 before publishing his book, and credit for the discovery is usually awarded to the English minister of religion and chemist Joseph Priestley (1733-1804).

Priestley discovered the same gas independently in 1774, and claimed that it might have therapeutic value for patients suffering from respiratory complaints. He believed firmly in the existence of 'phlogiston', the substance that was supposed to be the raw material of fire, and he called his gas 'dephlogisticated air'.

His discovery was widely reported and it led to his meeting the eminent French chemist Antoine Laurent Lavoisier (1743-94). Lavoisier was a member of the nobility and held a position in the prerevolutionary administration. This led to his arrest and, on 8 May 1794, he was tried along with 27 other people, and all of them were executed the same afternoon. The next day, a former colleague, Joseph Lagrange, commented that 'it required only a moment to sever that head, and perhaps a century will not be sufficient to produce another like it'.

Between 1775 and 1777, Lavoisier repeated the Priestley experiments and conducted more of his own. He called the element 'oxygine' (changed later to 'oxygène'), from the Greek *oxus* (acid) and *genes* (born), because he believed its principal role was in the formation of acids.

Lavoisier also claimed, however, that when oxygen combines with 'caloric' (an intangible substance that was believed to be the source of heat) it becomes the gas that had previously been known as 'pure air' or 'vital air'.

The commonest oxygen (O) molecule consists of two atoms bonded together, for which the symbol is O_2. Oxygen accounts for 20.94 per cent of the lower atmosphere, by volume. Oxygen atoms can become separated and then reform as molecules of three atoms rather than two. The symbol for this form of oxygen is O_3, and its common name is ozone. Ozone occurs mainly in the the upper atmosphere (see page 159), but there is a very small amount, 0.00006 per cent of the total volume, in the lower air — although the concentration varies from place to place and from time to time.

Scheele missed being given credit for the discovery of oxygen, but there is no doubt about his simultaneous (1772) discovery of nitrogen. This was the 'foul air' that inhibits combustion. He isolated it by removing 'fire air' from ordinary air by combining it with materials that burned or could be oxidized. The residue — nitrogen — would not support combustion. Priestley and Daniel Rutherford, a student of Joseph Black, made the same discovery at about the same time as Scheele — Priestley called it 'phlogisticated air'. Lavoisier identified the gas as an element, and found that organisms die when they are exposed to an atmosphere of the pure gas. The name 'nitrogen' was introduced in 1790, after the element had been found to be a constituent of nitre (potassium nitrate, or saltpetre).

Nitrogen (N) exists in the atmosphere as an almost inert gas, with a molecule composed of two bonded atoms and written as N_2. The gas accounts for 78 per cent of the lower atmosphere, by volume.

Together, nitrogen, oxygen and carbon dioxide add up to 99.05 per cent of the volume of the air we breathe. This leaves 0.95 per cent unaccounted for, and most of the 'missing' gas, 0.93 per cent of the total atmospheric volume, is another element, argon, with the chemical symbol Ar. Argon is colourless, odourless and tasteless, and it does not form true compounds with other elements. Its name, argon, is derived from *argos*, a Greek word meaning idle. Because argon is so inert, it is surprising that its existence was discovered as long ago as

1785, not long after oxygen and nitrogen had been identified, although the discovery was not recognized until more than a century later. Henry Cavendish (1731-1810), an English physicist and chemist, had already identified hydrogen (symbol H), which he called 'inflammable air'. Developing an experiment described by Priestley, he mixed hydrogen and air in a sealed flask, fired it with an electric spark, and obtained water (H_2O), but also some nitric acid (HNO_3), which he identified correctly. This demonstrated the compound nature of water. He then sought to discover whether nitrogen is a pure substance (an element) or a mixture of several substances. Again, he mixed hydrogen and air, this time over potassium carbonate (or potash, K_2CO_3) to absorb the nitrogen oxides, fired the mixture, removed surplus oxygen, and found he was left with a small remainder he could not remove by adding more pure oxygen and igniting it. Convinced he had removed all the nitrogen, he concluded that nitrogen is mixed with a small amount, about one part in 120, of some other substance. This must have been mainly argon.

It was not until 1894 that Lord Rayleigh (1842-1919), the English physicist and mathematician, and Sir William Ramsay (1852-1916), the Scottish chemist, rediscovered this observation by Cavendish. They had found that nitrogen collected from the air is very slightly denser than nitrogen obtained chemically. They isolated argon, for which Rayleigh was awarded the 1904 Nobel Prize for physics, and learned much about the physical and chemical properties of argon, and also of other inert (or noble) gases, for which Ramsay received the 1904 Nobel Prize for chemistry.

Argon is formed by the decay of potassium-40, the radioactive form (isotope) of potassium. This has a half-life (the time it takes for its radioactivity to be reduced by half) of 1.31 billion years, and the ratio of potassium to argon is used for dating rocks.

Having identified argon, Ramsay continued the experiments, trying to find other ways to obtain the gas. In 1895, he heated cleveite, a mineral that contains uranium, and obtained argon, but also another gas, helium. It, too, occurs in the lower atmosphere, a product of the radioactive decay of uranium, and accounts for 0.0005 per cent of the total volume. Its chemical symbol is He.

Argon and helium are both noble gases and, in the course of his studies of argon, in 1898 Ramsay, and his colleague Morris William Travers, discovered three more. They took the name of one from the Greek word *neos*, meaning 'new', and called it neon, with the symbol Ne. It accounts for 0.0018 per cent of the volume of the lower atmosphere. Krypton (symbol Kr), a name derived from the Greek *krupton*, meaning 'hidden', accounts for 0.00011 per cent of the volume of the atmosphere. Xenon (symbol Xe), from the Greek *xenos*, meaning 'strange', is even rarer — the air contains only 0.000008 per cent of it.

The only permanent constituents of air that remain are very small traces of hydrogen and methane (CH_4). Water vapour is not listed among the atmospheric ingredients, because its concentration is too variable.

Changes in air

Air consists of nitrogen and oxygen, with a small amount of carbon dioxide and traces of other gases, and these ingredients are present in the same proportions throughout the lower atmosphere (but not in the upper atmosphere, see page 57). Where variations occur, they are local and temporary.

We do perceive variations. The air may feel 'fresh', or 'stuffy', or we may be aware of a pollutant but, apart from pollution, such changes are due to an increase or decrease in the concentration of water vapour, or of suspended particles — dust — and water vapour and solid particles are not included in the list of chemical ingredients.

When the air feels 'stuffy', for example, it is because it is humid. Our bodies cannot cool themselves so easily by the evaporation of perspiration, and we feel uncomfortable. In a closed and very crowded room, it is possible that, because of the number of people breathing, the proportion of carbon dioxide may be elevated and that of oxygen lowered, but by far too small an extent for the human body to detect it. The 'stuffiness' is due to humidity — we exhale water vapour. 'Fresh' air is relatively dry and dust-free. The air we breathe is a mixture of 78 per cent nitrogen and 21 per cent oxygen regardless of whether we are inside a building, in a desert, or on the deck of a ship far out at sea.

This constancy in the composition of the air is a mark of its capacity for self-regulation. Perturbations can and do occur, but the balance is restored very quickly. Air is not an inert, passive substance.

Oxygen, its second largest ingredient, is one of the most reactive of elements. It combines readily with a wide variety of other elements — water, after all, is a result of its combination with hydrogen — and sometimes the reaction is rapid and 'exothermic' — it gives off heat. This is what happens when materials burn, and in an atmosphere containing such a high proportion of oxygen as that of Earth, fires start very easily. Forest, bush and grass fires occur quite naturally, most commonly ignited by lightning, and they are often extensive. As they burn, the gaseous products of their combustion — principally water vapour, carbon monoxide (CO), carbon dioxide, the oxides of nitrogen and sulphur dioxide (SO_2) — enter the air and, for a time at least, they become part of it.

Fires have little overall effect on the chemical composition of the atmosphere. This is not to say they are unimportant for other reasons — quite small changes in the concentration of carbon dioxide may produce climatic changes, for example — but only that the quantity of gas released by even a major fire is extremely small when compared with the total mass of the atmosphere, which is about 5 million billion (5×10^{15}) tons, half of it compressed between the surface and a height of about 3 miles (5 km) and, up to a height of about 8 miles (13 km), filling a volume of about 1.6 billion cubic miles (6.6 billion cu km).

Dramatic though they may be, fires form part of the cycle by which elements move between land and air. The carbon dioxide that is added to the air came from the air in the first place, when it was used in photosynthesis by green plants, so it is not so much being added as returned — though perhaps sooner than it would have been otherwise. Water vapour is produced in a fire by the combination of oxygen with the hydrogen contained in all living, or once-living, matter. Plant matter and fossil fuels (peat, coal, natural gas and petroleum), which were once living, are called 'hydrocarbons', and the word 'hydrocarbon' means 'hydrogen plus carbon'. Again, plants obtained the hydrogen from water, so the water vapour is being returned to the air rather than added to it.

(Opposite) Bush, forest and grass fires are common and often extensive. The water vapour and smoke they release into the air soon returns to the ground, but the oxidation of carbon in vegetable matter adds carbon dioxide. This fire is in north-eastern Brazil.

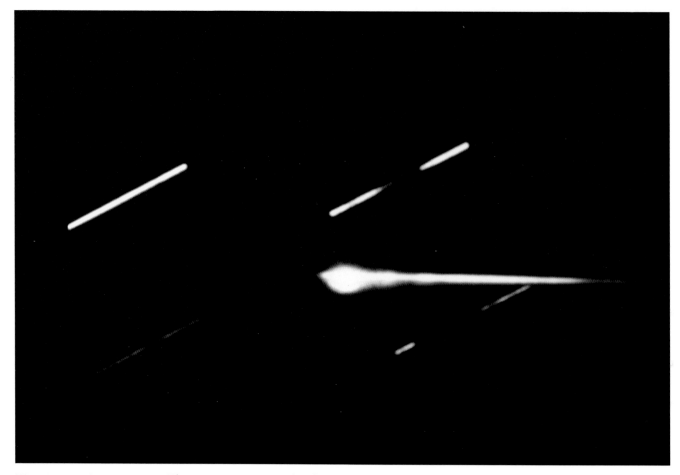

'Shooting stars' are particles that reach Earth from space and are vaporized by the heat of friction as they enter, and become part of, the atmosphere.

Elements move in cycles, but very small amounts are also added to the air, and lost from it, at a fairly constant rate. Volcanoes are the most important source of new gases, brought from deep below the surface as molten rock moves upwards and the great pressure under which it is contained relaxes. This causes certain ingredients of the rock — called 'volatiles' — to vaporize and escape, introducing new atmospheric material. Volcanoes emit a variety of gases, including water vapour (usually the main ingredient), carbon monoxide and dioxide, nitrogen, sulphur dioxide, hydrochloric acid (HCl) and compounds of fluorine and bromine. Most remain in the lower atmosphere, but hot gases and dust ejected in a violent volcanic eruption may be carried to a height of more than 12 miles (20 km), where winds distribute them over a very wide area.

Matter also enters from space, as meteors and meteorites, the great majority of which are no larger than a grain of sand and vaporize in the upper air. The 'shooting stars' that can sometimes be seen on a dark night are meteors, but most are invisible. They enter the atmosphere constantly at a rate of several tons a day.

The decay of radioactive elements produces a range of other elements, some of which are gases and may enter the air. Radon, for example, is an element (chemical symbol Rn) produced by the decay of another element, radium (Ra). It enters the air, but is not included in the list of atmospheric ingredients because it is very short-lived. Its most stable isotope (explained below) has a half-life of only 3.8 days. (The half-life of an unstable isotope is the time it takes for its radioactivity to be reduced by half; the radioactivity is effectively exhausted after 20 half-lives.)

Most natural elements exist as a mixture of atoms, in the nuclei of which the number of protons (particles carrying a positive electrical charge) is the same, but the number of neutrons (particles that are electrically neutral) varies. The number of protons determines the chemical characteristics of the atom, so the atoms are all chemically identical, but there are slight physical differences due to the variation in the number of neutrons. These different versions of otherwise similar atoms are called 'isotopes' of an element. Some

isotopes are unstable, which means they are radioactive, and decay to more stable isotopes. The decay may involve the loss of two protons and two neutrons together. This is called 'alpha decay' and, because protons are lost, the chemical behaviour of the atom changes and it is transmuted into a different element. Radium undergoes alpha decay to become radon, which also undergoes alpha decay, transmuting into a series of short-lived radium isotopes, then to one (called radium D) with a half-life of 20 years, which decays eventually to lead.

The nucleus of a helium atom consists of two protons bound to two neutrons. In other words, it is the particle emitted by an atom that undergoes alpha decay. Elements such as radium, uranium and thorium, which undergo alpha decay, are distributed throughout most of the rocks from which the Earth's crust is made, and so helium is being produced constantly. Most of it remains trapped by the rocks from which it forms, but it can collect in certain places and natural gas contains it, at concentrations of about 1 to 7.5 per cent by volume (and natural gas is the main source for the extraction of helium for industrial use).

The most common helium isotopes are stable, and helium does not form compounds with other elements, so you might suppose that over millions of years the gas would accumulate in the atmosphere. It does not accumulate, because helium is much lighter than nitrogen or oxygen. Its relative atomic mass is 4, compared with 14 for nitrogen and 16 for oxygen (this difference explains the value of helium as a lifting gas in balloons). As it mixes with the air, therefore, helium tends to rise. The ordinary movement of air causes such thorough mixing that the proportion of helium does not increase with height, except near the very top of the atmosphere — a place from which light atoms may escape from Earth into outer space.

The atoms and molecules of a gas move about freely and at random. Collisions are frequent. These cause no damage, but they deflect the particles, rather like the way moving balls are deflected when they collide with one another. Those which rebound upwards do not travel far before another collision bounces them sideways or downwards again.

At the very top of the atmosphere,

Volcanoes inject a range of gases into the air.

Farming is possible because nitrogen, an essential plant nutrient, is converted into soluble compounds that can enter the root system. (Opposite, top) Fires oxidize their fuel, removing oxygen from the air. (Opposite, bottom) Hematite is a common iron-oxide mineral formed when photosynthesizing organisms first began releasing oxygen. Oxygen could not accumulate in the air until the oxidation of iron compounds and other oxidizable substances was completed (see page 21).

however, the gas is very rare — its atoms and molecules are much more widely spaced and the greater the altitude the more widely spaced they become. A particle travelling upwards has significantly less chance of being deflected downwards again by a collision, and so some escape altogether. It is because the air has mass that the gravitational attraction of the Earth is able to retain it. To escape from this attraction, a body must travel directly away from the Earth at a minimum speed of 6.96 miles per second (11.2 km/s) — known as the 'escape velocity'. This speed (about 25,000 miles per hour — 40,225 km/s) is fast for most moving objects, but not for an atom in a gas.

Hydrogen (with a relative atomic mass of 1) is even lighter than helium. It also

escapes from the top of the atmosphere and it, too, is replaced, but at a great height rather than from the ground. Hydrogen is released by the oxidation of methane (CH_4) into carbon dioxide and water and then the breakdown of water.

The methane is a by-product of biological activity (see page 30). Like helium, it is released at a fairly constant rate, but its concentration rarely alters (although it is increasing at present because of certain human interventions, see page 157), because it cannot survive for long in the presence of oxygen.

Oxygen is reactive, but nitrogen is not, yet this gas is also removed from the air. Most of the removal is biological, and the nitrogen removed from the atmosphere by one group of living organisms is returned to it by another group, but some is also removed by purely physical and chemical means.

If a mixture of nitrogen and oxygen is heated strongly, or if a large amount of energy is applied to it in some other form, one molecule (two atoms) of oxygen will bond with one molecule (two atoms) of nitrogen to produce two molecules of nitric oxide (NO). The energy required is considerable — it takes about 181,000 joules (43,200 cal) to produce two molecules — but lightning, which is a big electric spark, provides enough and, although electrical storms may be uncommon events in any one place, over the world as a whole they are a daily occurrence. Intensely hot fires, some of which occur naturally, also generate temperatures high enough for nitric oxide to form.

In the air, nitric oxide combines rapidly with free oxygen to form nitrogen dioxide (NO_2), some molecules of which link together to form N_2O_4, and the mixture of NO_2 and N_2O_4 dissolves in water to become nitric acid (HNO_3). As nitric acid it is washed to the surface. In the soil it reacts to form nitrates, especially with potassium to form saltpetre (KNO_3) or (more commonly) with sodium to form Chile saltpetre ($NaNO_3$).

Gases are also removed from the air at ground level, rather than in the air itself. You will be aware of this if you have ever owned an old car — or any other article made from iron or steel (steel is made almost entirely from iron, of course) that you have kept outdoors. The metal rusts.

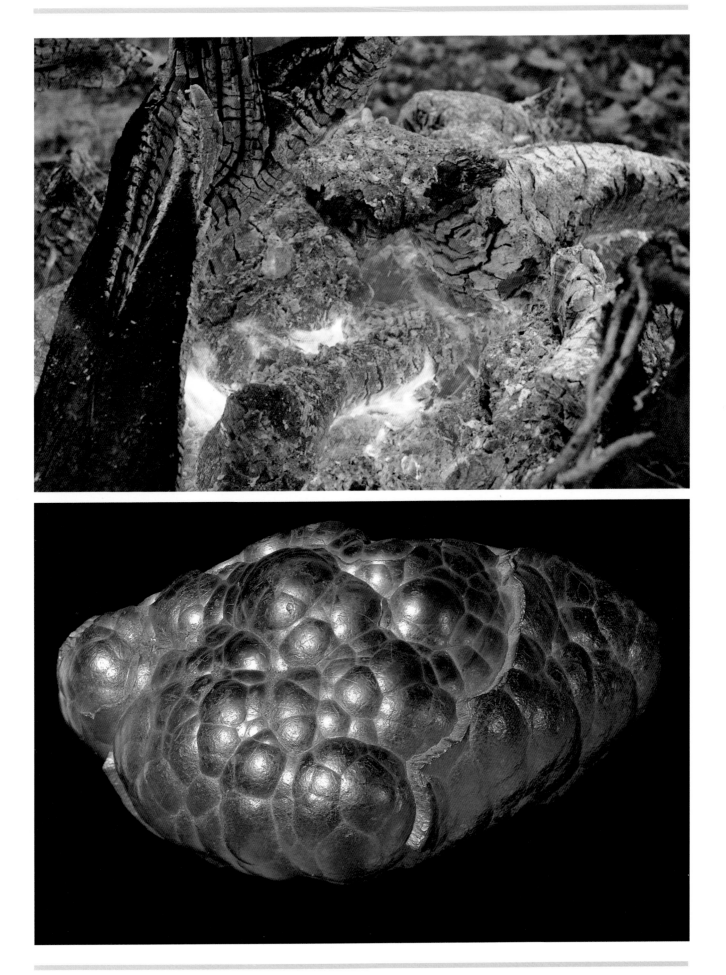

Were you to weigh a sample of iron very accurately, leave it exposed to the air until it was thoroughly rusted, and then to weigh it again, you would find that rusted iron weighs more than the original iron. The additional weight is supplied by oxygen, for the rust is an oxide of iron. Rusting cannot occur in very dry air. It is an electro-chemical process which requires water. The rather complex reaction proceeds in two stages, eventually to produce a hydrated oxide ($Fe_2O_9.2H_2O$).

The oxygen that is removed from the air by the rusting of iron is removed perma-nently, and iron is not the only metal that oxidizes readily — most metals do so. Aluminium, for example, does not occur naturally in its pure, metallic form. It must be obtained by refining. When pure alu-minium is exposed to the air, the outermost layer of its atoms oxidizes in a matter of seconds — and the metal itself is usually extracted from its ore, bauxite, which is an oxide (Al_2O_3).

Oxygen is lost from the air quite readily. We may regard this as a disadvantage when we see patches of rust appearing on cars, but there are other ways in which we benefit

greatly — because the loss of oxygen helps to keep the air clean. The extent to which a pollutant can damage living cells depends on its ability to form chemical bonds with molecules in those cells — its reactivity. While it is airborne, a pollutant molecule which encounters a molecule of oxygen is likely to bond to it. This changes the molecule into an oxide and, in most cases, reduces its capacity for entering into further chemical reactions, so it is rendered less harmful. At the same time, the addition of oxygen means the molecule becomes larger and heavier. This increases the likelihood that it will fall or be washed from the air by rain.

The atmosphere is not depleted by this loss of oxygen, because there is only a relatively small amount of oxidizable material exposed to it, and green plants are constantly adding oxygen to the air as a by-product of their photosynthesis. The loss does help prevent the concentration of oxygen from increasing and this, too, is advantageous. Were the air to contain much more oxygen than it does — say about 25 per cent instead of 21 per cent — hydrocar-bons would burn much more readily than

(Above) The energy of lightning causes the oxidation of nitrogen. The soluble oxide dissolves in rain droplets and is washed to the ground. (Opposite, top) Algal mats, like these at Sharks Bay, Western Australia, are communi-ties of bacteria and cyanobacteria. They resemble closely those found as fossil stromatolites which are believed to have released free oxygen into the air. (Opposite, bottom) A green sea contains large numbers of minute green plants, all releasing oxygen into the air. (*See* page 33 for the atmospheric importance of photosyn-thesis.)

As waterfalls (this one is in Teesdale, England) bubble and splash over the rocks, they entrain air, some of which dissolves in the water.

they do, and the resulting fires would cause great devastation. They would also consume oxygen, of course, and so reduce the atmospheric concentration.

Expose copper to air, and before long it may acquire a green coat of verdigris. The precise composition of this varies according to local conditions, but it usually includes copper carbonate $[CuCO_3.Cu(OH)_2]$, sulphate $[CuSO_4.Cu(OH)H_2O]$ and, if there is sea (or road) salt in the air, a chloride $[CuCl_2Cu(OH)_2]$. The exposure of copper, therefore, removes carbon dioxide, sulphur dioxide and chloride ions from the air. Silver also removes sulphur dioxide, which forms

the dark, silver-sulphide tarnish the metal accumulates if it is exposed to the air for any length of time. Polishing removes the tarnish, but it does not return the sulphur dioxide to the air.

Gases and solid particles which enter the air may be washed from it — often in a matter of hours or days. Sulphur dioxide, for example, is emitted from many volcanoes, and from the burning of most fuels, because sulphur is an essential nutrient for all living organisms, and all organic (living or formerly living) material contains at least some sulphur. Once in the air, sulphur dioxide (SO_2) either adheres to a solid surface (release it in a closed room and adhesion to surfaces will remove it from the air in a matter of minutes) or dissolves into water droplets to form sulphuric acid, so it falls to the surface as a constituent of rain or snow, or of the layer of moisture which mist deposits on any surface with which it makes contact. A molecule of sulphur dioxide will remain in the air for a few hours or, exceptionally, for up to one month.

Dust particles are washed from the air in the same way, and about as quickly. This is why air that has crossed an ocean contains very few solid particles. Salt — sodium chloride (NaCl) — also enters the air. Indeed, if you visit a rocky coast you can watch it do so. As the waves strike the rocks, spray is thrown high into the air, as a white mass of foam. The water is white because it contains many bubbles — thin coatings of water that are wrapped around tiny parcels of air. The water is sea water, of course, and saline. Most of the foam falls back into the sea, but some of it is thrown high enough for the water to evaporate while it remains airborne. When that happens, it is fresh (non-saline) water vapour which enters the air, leaving behind tiny crystals of salt. Sodium chloride is not a constituent of the air, however, because the tiny crystals — which are always found in very moist air, of course, because the water vapour which evaporated to produce them has not ceased to exist — attract water vapour. It condenses on to the crystals, dissolving them in the process, and they fall back to the surface, perhaps after a brief sojourn as part of a cloud.

Bubbles also have other effects. Oxygen and carbon dioxide are slightly soluble in water. The extent to which these gases will dissolve depends on the temperature, but they are never very soluble. At 41 °F (5 °C), water will be saturated with oxygen when it holds about 8.9 parts of oxygen to 1000 parts of water, and about 6.4 parts per thousand at 68 °F (20 °C). Oxygen dissolves into water at the surface where the two fluids meet, and still water may never reach saturation except near its surface. More oxygen will dissolve if the surface area of the water can be increased, and the most obvious way to achieve this is to agitate the water violently, so it splashes, throwing drops so high into the air that they can be saturated before they fall, and forming bubbles that can absorb oxygen even more

rapidly, because their 'skin' of water is exposed to the air on the inside as well as the outside. Turbulent, fast-flowing streams, with rapids and waterfalls, usually carry water that is saturated with oxygen, for this reason.

Aquatic organisms use the oxygen that is dissolved in water for respiration. This process involves adding oxygen to other compounds and the end products are adenosine triphosphate (ATP, a compound that delivers energy to cells and tissues and remains within the body), carbon dioxide and water (which is why the breath we exhale is moist). The oxygen is removed from the air very effectively, but eventually it is returned by green plants as a by-product of photosynthesis, so the balance is maintained.

There is a mechanism by which carbon dioxide is removed from the air more or less permanently. On Earth, living organisms have accelerated the process to such an extent that many scientists maintain they have taken it over, so it has come to be essentially biological, but it would still proceed, albeit much more slowly, even if our planet were lifeless.

Carbon dioxide is only slightly soluble in water and, as with oxygen, its solubility decreases as the temperature rises. At 32 °F (0 °C), saturation is reached at about 3.3 parts of carbon dioxide to 1000 parts of water, and at about half of that concentration at 68 °F (20 °C). This is very little, but the amount of available water is extremely large.

When it dissolves, some of the carbon dioxide forms carbonic acid (H_2CO_3), and the remainder bicarbonate (HCO_3) ions, and the reactions proceed in either direction, so carbonic acid and bicarbonate are constantly dissociating and reforming. When water containing carbonic acid, and saturated with carbon dioxide, comes into contact with rocks containing calcium, the two react to form calcium bicarbonate [$Ca(HCO_3)_2$], which is soluble. If the water then mixes with water containing little dissolved carbon dioxide, or is exposed to the air, the calcium bicarbonate will dissociate, yielding carbon dioxide, water and calcium carbonate ($CaCO_3$) — which is insoluble and settles as a precipitate.

In shallow seas, which is where rivers discharge their waters, this leads to an accumulation of calcium carbonate sedi-

ment. Very slowly, over periods measured in hundreds of millions of years, the movements of continents carry the sediment to regions where it may be raised above the surface, as newly formed mountain chains, or subducted back into the Earth's mantle, below the crust. The carbon dioxide, contained in the calcium carbonate, is buried and out of circulation. It will return eventually, through the erosion of rocks made from sediments that have been returned to the surface, or from volcanoes, but, in the history of the Earth, the cycle has removed more carbon dioxide from the air than has been returned to it.

Calcium carbonate is insoluble in surface water, but it becomes increasingly soluble at lower temperatures and at higher pressures it dissociates, releasing carbon dioxide. These conditions occur in the deep oceans, where there is a 'carbonate compensation depth' below which carbonates break down to release carbon dioxide faster than they form, so that the insoluble calcium carbonate sinking from above fails to reach the ocean floor and there is no accumulation of sediment. In the Pacific Ocean, the carbonate compensation depth lies between about

13,000 and 16,500 feet (4000-5000 m). The release of carbon dioxide below the carbonate compensation depth does not return it to the atmosphere, however, because there is little mixing between deep water and surface water — the deep water is colder and, therefore, denser than the overlying water.

On Earth, the composition of the atmosphere is now controlled almost wholly by the activities of living organisms (see pages 33-44) but, in their absence, gases and solid particles would continue to enter the air and to leave it. The biological contribution accelerates the cycling of atmospheric constituents. It is not possible to say precisely how rapidly molecules move between the air and surface, but rough estimates have been made of the average length of time a molecule or particle is likely to remain in the air — its 'residence time'. Argon stays in the air for ever. The residence times for other ingredients are: nitrogen 10 years to 1 million years; oxygen 1 year to 1000 years; carbon dioxide 1 month to 100 years; sulphur dioxide 1 minute to 1 month; nitric oxide, chloride (from sea salt), and mineral particles 1 minute to 2 months.

(Above) Clouds form when water vapour condenses on to small solid particles. The water then dissolves soluble gases and the rain washes them, and the particles, to the ground. (Opposite) When the sea freezes, the water beneath the ice becomes very dense and sinks, taking with it carbon dioxide and oxygen dissolved from the air.

HOW DID THE ATMOSPHERE EVOLVE?

'I feel confident I can carry out the work of the facade of San Lorenzo in such a way that it will be the architectural and sculptural mirror of all Italy ... I have ordered many marbles ... and I have started quarrying in various spots. And in some places where I have spent good money, the marbles did not turn out the way I expected, for marbles are unpredictable, especially when it comes to the large blocks I need, and which I want to be as beautiful as possible.'

San Lorenzo was to be the family church of the Medici, in Florence, and Michelangelo wrote the letter from which this extract is taken in April, 1517. He was writing from Carrara, in Tuscany, which is where he obtained the pure white, hard marble from which he made his finest work. Not all marbles are white. Siena marble, also quarried in Tuscany, has red mottling and, in other places, chemical changes or impurities in the stone can produce grey, brown, yellow or green pigmentation.

Sculptors have always loved marble. Indeed, the word itself is derived from the Greek *marmaros*, which means, simply, 'shining stone'. It occurs all over the world. The roof of the Lincoln Memorial, in Washington, includes marble from Yule, Colorado, and the marble used for the statue of Abraham Lincoln came from Georgia. Not far away, the facing on the National Gallery of Art is made from marble quarried in Tennessee.

Marble is formed from limestone that has been subjected to high temperatures and pressures, which have radically altered its crystalline structure — the process is called 'metamorphism'. This transformation produces a stone that can be cut and shaped, but which is hard enough to be polished. Altered though it is, however, marble remains a form of limestone, and limestone consists mainly of calcite or dolomite. Calcite is calcium carbonate ($CaCO_3$) and dolomite is calcium-magnesium carbonate [$CaMg(CO_3)_2$]. 'Limestone' is usually defined as a rock consisting of at least 50 per cent calcium carbonate.

If marble occurs widely and is plentiful — the Carrara quarries were opened in about 40 BC and they are still being worked — limestone is very common indeed.

Chalk, another form of calcium carbonate, is hardly less common. It underlies much of the rolling hills of southern England (the Downs), and is exposed in the dramatic White Cliffs of Dover (which also occur in northern France). In other places, where hills made from horizontal beds of limestone have eroded, 'limestone pavements' lie exposed. These regions, where a bare rock surface is intersected by deep fissures ('grikes'), which trap enough soil for plants to grow, are often extensive. Like other limestone landscapes, they are found in many parts of the world.

As a class of rocks, limestones and chalks are quite easy to identify. All you need is a small bottle of dilute hydrochloric acid. Sprinkle a few drops of the acid on to the rock and, if it fizzes, the rock is 'calcareous' (the word means 'made from or containing calcium carbonate') — a limestone. The fizzing is caused by a chemical reaction between the acid (HCl) and the calcium carbonate ($CaCO_3$), that produces calcium chloride ($CaCl_2$), water (H_2O) and carbon dioxide (CO_2). The mixture fizzes as the carbon dioxide bubbles through the water.

A cursory examination of most marbles can provide little more information than this, but look closely at a formation of carbonate rocks which have not undergone metamorphism, and you may be able to see that they appear to lie in bands. These may not be arranged horizontally and may not be straight, but that is because they have been deformed by movements in the Earth's crust. Picture them straightened out and flat, and you will have an idea of the way they began — as sediment on the bottom of a sea. Look closely at them and you may be lucky enough to find fossils of what were once marine animals.

We know that carbon dioxide can be lost from the air, and that its removal is more or less permanent when it reacts in water with calcium to form calcium carbonate, which is insoluble. The calcium carbonate settles to the sea bed, where it accumulates. Sediments, which begin as mud, can be altered into rock by pressure and heating. This is how sedimentary rocks form, and limestones and chalks are sedimentary rocks — identifiable as such partly by their layering, which records the way they accumulated. It is difficult to avoid the conclusion that it was the removal of carbon dioxide from the air, and calcium from rocks over which rivers flowed, that supplied the raw materials from which calcareous sedimentary rocks were made.

This leads to a further conclusion. Because calcareous rocks are so abundant, taken together they represent the removal of a truly vast quantity of atmospheric carbon dioxide. Today, carbon dioxide accounts for only about 0.03 per cent of the volume of the atmosphere but, if all the carbon dioxide bound in the sedimentary rocks were returned to the air, the proportion of carbon dioxide would increase greatly. Indeed, the chemical composition of the atmosphere would be changed beyond recognition.

If the carbon dioxide now trapped in the rocks was taken from the air, then at one time it must have been in the air. It follows, therefore, that our atmosphere has not always been as it is today. It has changed over many years. The air has a history of its own.

Our atmosphere has evolved from an earlier and quite different atmosphere. Apart from a continuing removal of carbon dioxide, the composition of the air has not changed fundamentally over the last 600 million years, however, so it makes sense to think of the two atmospheres separately.

In addition to these, the Earth probably had yet another atmosphere, its first, for a short time after the planet formed. The Sun began as a cloud of gas and dust, which collapsed under its own gravitational attraction. Almost all of the matter became part of the new star — the Sun itself — but a small amount remained in orbit about the central mass, and it is from this residue that the planets formed.

Hydrogen and helium were the most abundant ingredients of the original cloud — they are the principal constituents of the Sun — and also the lightest. It seems likely, therefore, that the newly formed Earth may have had an atmosphere composed mainly of hydrogen and helium, possibly mixed with traces of other gases. The existence and composition of this very early atmosphere can only be inferred from the most plausible theories about the way planets form, for it vanished long ago and left no trace of itself. The Earth's own gravitational

attraction is strong enough to retain gases heavier than hydrogen and helium, but an atmosphere composed of those gases would have leaked away into space. If it existed, the hydrogen-helium atmosphere was probably lost fairly rapidly as pressures within the Sun reached the level at which atomic nuclei began to fuse together, and the star began to shine. Its warmth would have imparted enough energy to hydrogen and helium for them to have been swept away completely.

If this is what happened, the lost atmosphere was replaced very quickly by one that proved more substantial, and of this there is clear evidence. The decay of radioactive elements proceeds at a very precise rate, and allows scientists to discover the age of materials by analysing the isotopes from which they are made. Radioactive dating has established that the formation of the Earth was completed about 4.6 billion years ago. The oldest rocks so far discovered (in western Greenland) are 3.8 billion years old, and they are sedimentary. This means they formed by the accumulation of sediment, a process that can occur only in water. Liquid water must have been present while the rocks were being deposited and, if water was present, there must also have been an atmosphere. This is because the temperature at which water boils is inversely proportional to the atmospheric pressure. If the air pressure is zero, water cannot exist at all as a liquid — all of it will vaporize. Had this

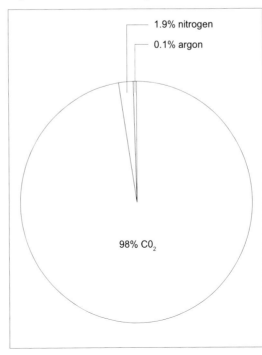

Supposed composition of Archaean atmosphere, by volume (see page 29).

happened, it is almost certain that the hydrogen and oxygen of the water molecules would have dissociated and been lost from the top of the atmosphere. Not only would the Earth have lost its atmosphere, it would also have lost all of its water.

It is quite certain, therefore, that the Earth had an atmosphere at least 3.8 billion years ago. That much we know, but our understanding of its composition is largely conjectural, although we are not quite helpless. The atmospheric gases must have been emitted by volcanoes — there seems no other way they can have originated and the Earth was much more volcanically active then than it is now. At that time more heat was being generated by radioactive decay beneath the crust, and this is the principal cause of volcanism. Less heat is generated now because the process has continued for long enough for much of the original radioactivity to have decayed.

Living organisms were already present 3.8 billion years ago — chemical evidence of them (not fossils) has been found in the Greenland rocks. They lived in water, because the evidence comes from sedimentary rocks, and they must have required particular chemical nutrients, especially carbon, nitrogen, sulphur and perhaps iron, all of which must have been present in the water, and some of which are most likely to have reached the water from the air.

Volcanoes release water vapour, hydrogen, carbon monoxide, carbon dioxide, hydrogen sulphide (H_2S), hydrochloric acid, nitrogen and sulphur dioxide, as well as compounds of bromine and boron. As all these gases mixed together and were exposed to sunlight they would have reacted with one another. Methane (CH_4) and ammonia (NH_3) would have been among the products, and those which are soluble, such as ammonia, hydrochloric acid and the sulphuric acid formed from sulphur dioxide, would have been washed to the surface as the water vapour condensed and fell as rain. Had there been any free oxygen, the hydrogen would have joined it to form more water vapour — and hydrogen that was not 'captured' in this way would have escaped into space.

Stars like the Sun grow hotter as they age. Scientists have calculated that about 4 billion years ago the Sun was radiating at about 70 to 75 per cent of its present intensity, but the temperature at the surface

was not much different from the temperature today. In fact, some scientists have estimated that it may have been warmer, at an average 73.4 °F (23 °C) — much like the present-day tropics. Certainly, the oceans did not boil, but neither did they freeze solid, for that would have inhibited the development of life and sedimentary processes. This implies some means by which the warmth reaching the Earth from the Sun was retained — in other words, a 'greenhouse effect'.

A 'greenhouse effect' (see page 161 for a fuller explanation) occurs when the atmosphere contains gases that are transparent to short-wave radiation, but partially opaque to long-wave radiation. The Sun radiates most strongly in the short-wave bands. This

warms the surface of the Earth, which then emits long-wave radiation. The overall effect is to allow energy to reach the surface unimpeded, but to delay its departure. There are many effective 'greenhouse gases'. They include water vapour, carbon dioxide, ammonia and methane — ingredients of the early atmosphere.

Most scientists now accept that a 'green-house effect' prevented the Earth from being extremely cold — with an average tempera-ture below freezing, and perhaps far below. This suggests that the principal constituent of the atmosphere must have been carbon dioxide. There were also traces of the noble gases, mainly argon, remaining from the earlier atmosphere and, perhaps, a little free nitrogen, although this is uncertain. Water

vapour must have been present (not usually included as one of the ingredients of the atmosphere), and there may possibly have been a very small amount of methane as well as a little ammonia formed by the reaction between hydrogen and nitrogen.

At this stage, the atmosphere contained, at most, the smallest traces of free oxygen, and probably it contained no oxygen at all. This means there was no formation of ozone (O_3), and no ozone layer to provide a shield against ultra-violet (UV) radiation — although UV radiation is much less biologi-cally damaging than many people suppose, and exposure to it is as likely to have accelerated the chemical reactions with which evolution began as to have hindered the process.

The Seven Sisters form part of the famous chalk cliffs of southern England. They are made from calcium carbonate, using carbon dioxide that was once part of the atmosphere.

There may have been a different kind of shield to filter incoming ultraviolet (UV) radiation, however. The UV radiation itself may have supplied the energy to drive chemical reactions between methane and water molecules, especially if the water molecules (H_2O) were broken into hydrogen (H) and free hydroxyl (OH) radicals, which are very reactive indeed. Such a reaction would have occurred in a region where the atmosphere was dense enough to absorb UV — which means it must have been at a high altitude — and it would have yielded a range of chemical compounds, some of them as liquid droplets and some as solid particles. In fact, it would have been very similar to a modern photochemical smog — the kind of pollution that forms over some cities — only at a high level rather than close to the ground. The atmosphere was what chemists call 'reducing', which means that more of its constituent molecules were in a state where they had a surplus of (negatively charged) electrons than were able to accept electrons. The mixture would have been chemically stable and, had it remained unperturbed, there is no reason to suppose it would ever have undergone any significant change. It was perturbed, however, when the evolution of an increasing variety of large, complex molecules produced the one we know as 'chlorophyll', and it became incorporated in the cells of the minute organisms that were the ancestors of the chloroplasts which today are found in the cells of plants. Photosynthesis had begun and probably it was well established by about 2.5 billion years ago.

Photosynthesis

Photosynthesis is a series of chemical reactions, powered by light energy (hence 'photo'), in which carbohydrates are synthesized from carbon, hydrogen and oxygen. The first organisms to exploit it were bacteria which obtain their hydrogen mainly from hydrogen sulphide (H_2S) or organic acids, and their carbon and oxygen from carbon dioxide (CO_2). The bacteria lived in water (as their descendants do to this day), and obtained the raw materials for photosynthesis from the water. Their removal of dissolved carbon dioxide allowed more carbon dioxide to dissolve into the water from the air. It may be that carbon dioxide was already being removed from the air by

forming insoluble carbonates and by dissolving into cold, saline, surface water that sinks to a great depth where it remains for a very long time. The advent of photosynthesis would have accelerated the process.

Bacterial photosynthesis was followed by another version of the process, using a different kind of chlorophyll, and it was this type of photosynthesis that led to the evolution of green plants. The essential difference is that in plant photosynthesis the hydrogen is obtained, not from hydrogen sulphide, but from the photolysis (breakdown using light energy) of water. Only the hydrogen is needed, and the oxygen is released as a waste product. The release of oxygen represented a profound chemical change in the strongly reducing environment.

The oxygen atom is able to accept electrons. 'Oxidation' means the loss of electrons and 'reduction' means the gain of electrons. Oxygen is able to oxidize other substances, and is reduced by doing so. At first, and for a long time, the oxygen that was released by photosynthesis was consumed by oxidizing the reduced molecules around it.

There is evidence for this. The oxygen combined with a number of metals to produce what are now important ores (see page 17). The commonest is hematite (which is sometimes spelled haematite, and is also known as iron glance, kidney ore, red iron ore, and specularite). It is an oxide of iron (Fe_2O_3), reddish, grey or black in colour and, if you know where to look, you can find large lumps of it. The most economically important deposits — mined in Western Australia, near Lake Superior in the United States, Labrador in Canada, the Ukraine and Brazil — formed as sediments on the floors of shallow seas from reaction with the oxygen that was being released by the early photosynthesizers.

Some of the oxygen entered the atmosphere and, because of the increase in bacterial activity, more methane was probably produced and it, too, entered the air. Bacterial activity is likely to have released nitrogen which, because it is relatively inert, simply accumulated as a gas. Meanwhile, the release of free oxygen was leading to the oxidation of some of the atmospheric gases. Methane, in particular, is oxidized by a series of steps to carbon dioxide and water. We exploit this reaction

(Opposite) Limestone pavement occurs in many parts of the world. This pavement, in the Yorkshire Dales, England, formed about 300 million years ago from the accumulated remains of marine organisms. Subsequent uplifting and erosion has exposed it.

when we use natural gas — which is mainly methane — as a fuel. A mixture of methane and oxygen will ignite readily and, when it does, the oxidation proceeds rapidly. In the more dilute atmospheric mixture the reaction would have been slower.

Photosynthesis was a great evolutionary success. It spread rapidly and, as it did so, the output of oxygen increased. It reached a point at which the methane content of the air fell sharply — oxygen was being produced more abundantly than methane, so the methane was being oxidized faster than it could be replenished. The oxygen would also have trapped any hydrogen that was released at the surface or from volcanoes. Hydrogen and oxygen react rapidly to form water — so the reaction would have augmented the quantity of water on the planet.

Then, as the supply of oxidizable compounds started to diminish, oxygen itself began to accumulate in the air. Soon an ozone layer must have formed, and then a point was reached at which the chemical character of the entire atmosphere changed, perhaps fairly abruptly. What had been a reducing atmosphere became an oxidizing atmosphere. This change was caused biologically, and had profound biological consequences, because the organisms that were living at the time had evolved in a reducing environment, with no free oxygen. So far as they were concerned, oxygen was a dangerous poison, the photosynthesizers polluters, and there must have been many deaths and not a few extinctions among the single-celled organisms.

The photosynthesizers increased in number, using carbon taken from the air to build the material of their own cells. It was a once-and-for-all expansion, which stabilized when the total mass of living organisms (the 'biomass') reached a certain level, and it involved the removal of a very large amount of carbon dioxide from the air. After that episode, the removal of carbon dioxide became more episodic and sometimes reversed, when the gas was released faster than it was being removed. As carbon dioxide was removed, methane destroyed, and oxygen and nitrogen accumulated around one billion years ago, the atmosphere began to resemble the one we have today.

If you visit Shark Bay, in Western Australia, or Laguna Figueroa, in Baja California, Mexico, you will see curious 'algal mats' (see page 18) projecting just above the surface of the shallow, extremely saline water. Irregular in shape, usually flat or slightly domed at the top, you might dismiss them with no more than a glance, thinking they are merely rocks, partly buried by sand or silt and with a superficial covering of some novel kind of seaweed. There are no rocks at their centres, however, and the mats are a great deal more interesting, and more important, than they look. The air we breathe, and the fact that we are able to breathe it, owe much to these puzzling and apparently unimpressive structures.

Far from being heaps of inert rock, sand or mud with simple plants growing on the surface, they are complex communities of cyanobacteria (blue-green algae), bacteria and other single-celled organisms. The mats trap silt from the moving water and produce calcium carbonate. The mats grow in layers, each of which has its own particular group of species and, at the surface, the cyanobacteria manufacture carbohydrates by photosynthesis.

The algal mats are alive today, but scientific interest in them centres on a strong suspicion that they may be modern examples of the most ancient communities on Earth, still existing as they have done for millions upon millions of years. Some of the oldest of all fossils, dating from about 2.3 billion years ago, are of structures called 'stromatolites', which closely resemble the contemporary algal mats. They are believed to have been communities of very simple, single-celled organisms, almost certainly dominated by cyanobacteria, which projected above the surface of shallow water. Locally, stromatolite fossils are common and stromatolite communities constructed the earliest of all reefs out of the limestone they formed — long before corals began to do so. Because they engaged in photosynthesis, the stromatolites released oxygen into the air and removed carbon dioxide from it, as did the green algae which are believed to have been very abundant at that time in surface waters, although they have left no fossils. The fossils are well known, but it is rare for organisms to be preserved as fossils. Traces of most of the stromatolites must have vanished over the very long period that has elapsed since they were alive. Those that have been found are only a small sample of a form of life that must once have been widespread.

(Opposite) Far from being inert, a fertile soil teems with life, its organisms continuing, but on a much larger scale, the cycling of gases begun by the algal mats. Surface plants engage in photosynthesis and, when plants and plant-eating animals die, the carbon in their remains is eaten by small animals and passed on until, eventually, it is released as carbon dioxide by decomposition, performed mainly by bacteria. These are some of the larger inhabitants of the soil — centipedes, millipedes, earwigs, woodlice and springtails.

About 2.3 billion years ago, when the stromatolites were flourishing, the climate of the Earth became colder. Indeed, the planet may have experienced its first ice age. The cooling seems to have coincided with the great expansion in photosynthesizing organisms and the release of free oxygen into the atmosphere. The increase in photosynthesis removed carbon dioxide from the air and free oxygen oxidized the methane. This reduced the strength of the 'greenhouse effect'. Different gases absorb radiation at different wavelengths (see page 163), which is why changes in the amount of methane are significant. Methane, absorbing at one wavelength, oxidizes to carbon dioxide and water vapour. These absorb at different wavelengths — from each other as well as from methane — but, if they are already abundant, adding to the concentration of them has little effect. The 'greenhouse gases' are sometimes likened to a blanket and, in this case, the comparison is apt. If you are lying in the cold with no blankets, a single blanket will help a great deal but, if you are already covered with enough blankets to keep you warm, adding another

will make only a slight difference.

This is a plausible explanation for that early cooling, but no one can say whether or not it is the true explanation, and there are alternative possibilities. The drift of continents might have carried a large land mass into polar regions. This might have increased the area covered by ice and snow and, therefore, the planetary albedo (see page 148), which would have produced a general cooling. Or the intensity of solar radiation might have fallen for a time — it is not constant, and its changes do affect the global climate.

If the cooling was due to the removal of atmospheric carbon dioxide and methane, however, it is the earliest example we have of climatic modification resulting from the activity of living organisms. This type of biological 'management' of the composition of the atmosphere, and hence of the climate, has continued, and it still operates today.

All of the oxygen in the air is the by-product of photosynthesis by green plants. We see them all around us — the grasses, herbs, shrubs and trees — but about one-third of all the oxygen is released by marine

The fossil remains of the 'skeleton' of a Devonian coral, *Hexagonaria percarinata*, made, perhaps 400 million years ago, from calcium carbonate using carbon dioxide removed, more or less permanently, from the air.

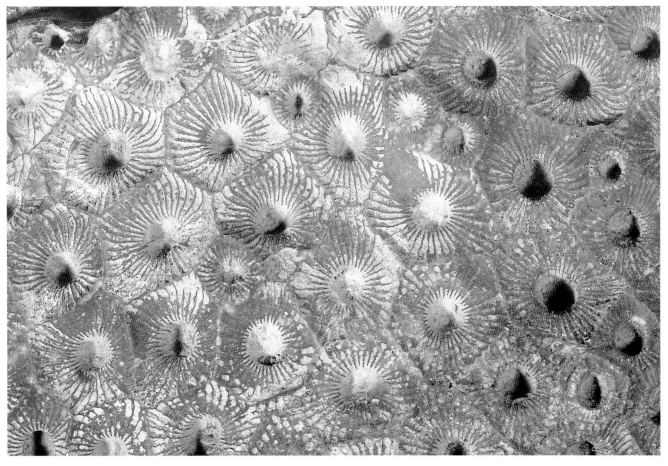

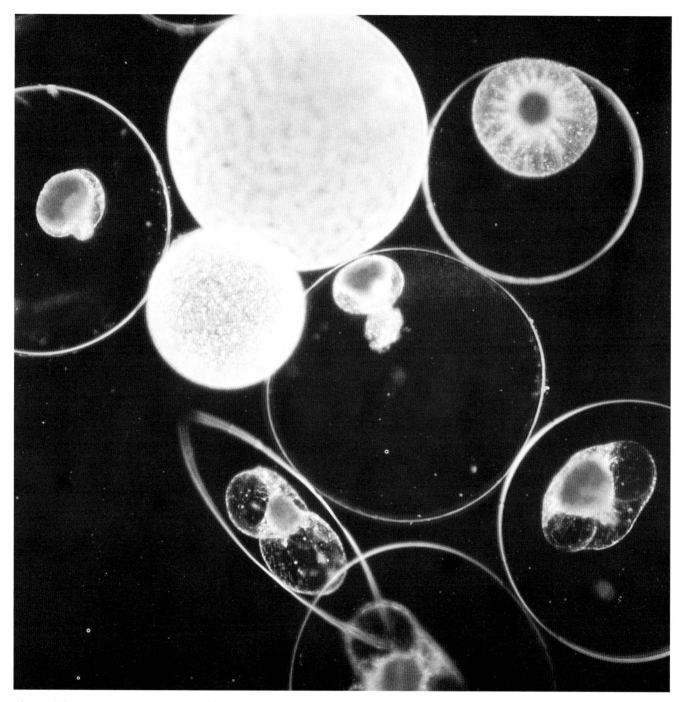

plants. These are mainly single-celled algae which drift near the ocean surface. They are known collectively as 'phytoplankton', and it was principally their ancestors — because they became abundant millions of years before plants established themselves on dry land — which raised the oxygen content of the atmosphere to its present level. It is these single-celled plants which colour the sea green (see page 18), and a blue sea is one from which they are absent — and consequently it supports few fish or other animals. The phytoplankton are not especially efficient photosynthesizers — area for area

the oceans are hardly more productive than deserts — but the oceans are immensely large. Atmospheric oxygen, then, is a wholly biological product.

Photosynthesis removes carbon dioxide from the air, but the green plants are not alone in this task. If you examine limestone or chalk rocks, you are quite likely to find fossils embedded in them. These are the remains or traces of once-living animals or plants and, although it is rare for them to be preserved in a recognizable form, they are not present by accident. The rock is largely made from them. Sand may also contain a

Plankton, the community of tiny plants and animals that drift near the sea surface, includes many single-celled organisms that are strikingly beautiful. The two opaque cells are dinoflagellates.

Chalk and limestone are made largely from the crushed, compressed and heated remains of the shells of once-living organisms, some of which are still recognizable as fossils. Chemically, the rocks and their fossils are calcium or magnesium carbonate and the carbon dioxide for the carbonate came originally from the air. The great abundance of chalk and limestone rocks can have formed only by the removal of truly immense amounts of carbon dioxide from the atmosphere.

large proportion of calcium carbonate. In some places, farmers use beach sand to add lime to their otherwise acid soils.

The reactions by which bicarbonate, supplied by carbon dioxide dissolved in water, and calcium, washed from rocks by rivers, form calcium carbonate are exploited by marine organisms. Because it is insoluble in shallow waters, calcium carbonate is an excellent raw material for the construction of shells. When the occupant of a shell dies, the soft parts of the body are eaten or decompose, but the insoluble, inedible shell sinks to the sea bed, where the shells accumulate, eventually to be compressed and heated by movements in the Earth's crust. That is how sedimentary carbonate rocks are formed, and the biological contribution far exceeds that of the inorganic, purely chemical reaction.

On a sandy beach that is littered with sea shells, the sand is likely to be calcareous. If you place a pinch of the sand on the palm of your hand and examine it through a hand lens, you will see that the sand consists of a mixture of quartz sand grains and tiny shell fragments.

Once converted into calcium carbonate, carbon dioxide has been removed from the air for a very long time indeed. It can be returned by exposing the carbonate to acidic water, or by heating it strongly — which drives off the carbon dioxide as a gas, leaving calcium oxide, or 'quicklime' — but most remains chemically immobilized for at least many millions of years. The sea-bed sediment becomes chalk or limestone, and eventually sand may turn into sandstone. Some of the sedimentary rock may eventually be exposed on dry land, but some is carried to regions where one of the plates, from which the Earth's crust is formed, is being carried beneath its neighbour ('subducted'), and its covering of sedimentary rock is carried with it, to be absorbed into the mantle lying below the crust. This does not necessarily remove the carbon dioxide from the air for ever. Some of the carbon dioxide emitted by volcanoes is derived from subducted sedimentary rocks.

Whole shells are rarely preserved, partly because the great majority are smashed against one another by the movement of the water and then crushed by the weight of those lying above, and partly because most of them were not really 'shells' in the first place. They were microscopically small plates that covered and protected single-celled organisms. The most numerous of these organisms are algae — plants, although they are able to move about as animals do — known as coccolithophorids. Their calcareous plates are called 'coccoliths'. Coccolithophorids became very abundant some 200 million years ago and have remained so ever since. Chalk is made mainly from their remains — including the White Cliffs of Dover, which were formed during the Cretaceous Period, which lasted from about 144 million years ago to 65 million years ago.

The growth of plants, and of the animals which feed on them, maintains a constant balance between the proportions of oxygen and carbon dioxide in the air. During their lives, organisms build their tissues from carbon, which is taken initially from the air by plants. When they die, their tissues decompose. Decomposition is mainly a process in which the carbon in tissues is oxidized, removing oxygen from the air and replacing it with carbon dioxide. The amounts of oxygen and carbon dioxide taken from the air and returned to it are equal. What might happen, though, were some catastrophe to exterminate all plant life? Animal life would also disappear, of course, because animals obtain their food from plants, but would the oxygen disappear from the air? This semi-permanent storage of carbon dioxide supplies the answer. The death of all plants would be followed by their decomposition, in which the carbon in their tissues would be oxidized, so the proportion of oxygen in the atmosphere would decrease and that of carbon dioxide would increase, but oxygen would not disappear. The chalk and limestone rocks are made from carbon dioxide that would not be returned to the air.

The fossil fuels

The so-called 'fossil fuels' — peat, coal, petroleum and natural gas — are also made from carbon that has been removed from the air by plants and then stored. Petroleum and gas contain no fossils, of course, because they have been formed under extreme conditions that have destroyed every trace of the original forms of the organisms from which they were made. Identifiable fossils are often found in coal, however, and peat, which has formed more recently and under less extreme conditions, contains plant

Petroleum is made from the remains of marine organisms that have been subjected to heat and pressure under airless conditions. This oil well is in Kuwait.

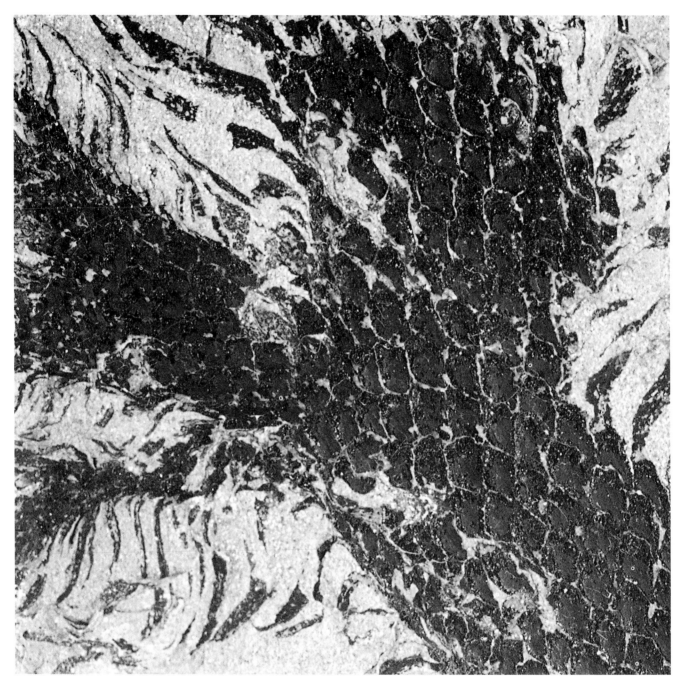

Coal consists of the partly decomposed remains of plants that lived many millions of years ago. It is not surprising, therefore, that it contains many fossils. This fossil was once part of a lepidodendron, a tree that grew up to 100 feet (30 m) tall.

remains that are easily seen. Sometimes it is possible even to identify their species.

Peat and coal are the remains of vegetation that has died but has been only partially decomposed, because the environment into which it fell was too airless, or too acidic, to be tolerable for the decomposer organisms — mainly bacteria. Peat is still being formed, although much more slowly than it is being used, and the peat which is cut for fuel is, at most, thousands of years old. The age of coal is measured in hundreds of millions of years. Some of the plants from which it is made were living about 400 million years ago, during the Silurian Period

(about 438-408 million years ago), but most lived during the Carboniferous Period (about 360-286 million years ago). The carbon dioxide that was removed from the air to make the hydrocarbons of peat, coal, oil and gas has been stored, but much less securely than the carbon dioxide which is 'locked up' in the carbonate rocks. The material is combustible and, when we burn it, the carbon dioxide is returned to the air.

The plants that were transformed into coal grew in swampy areas. When they died, they fell into muddy water and were covered by wet mud from which air was excluded. Conditions like these still exist today —

they have always existed — and, although no air-breathing organism is able to live in them, they are far from lifeless. Evolution began, and continued for a long time, under oxygen-free conditions. The accumulation of free oxygen, which altered the composition and chemical behaviour of the atmosphere, presented a severe challenge to the cells that were then living. For them, oxygen was extremely poisonous. Many must have died, but some survived in environments that protected them from this dangerous gas. Densely packed, waterlogged mud was one such environment and, much later, the digestive systems of animals (including humans) provided another. These are places inhabited by bacteria that have never adapted to the presence of oxygen. They continue to live in their ancient fashion and, as a by-product of their chemistry, some of them produce methane — a minor (but increasing) constituent of the atmosphere.

Bacteria

Many bacteria obtain the energy they require from organic matter, which they help to decompose, different bacteria specializing in particular nutrient elements.

If free oxygen is present, this decomposition involves chemical reactions which lead to the oxidation of carbon to carbon dioxide, of nitrogen in the form of ammonium (NH_4) compounds to nitrate (NO_3) and of sulphur compounds to sulphate (SO_4). Where there is no free oxygen, energy is provided by different (and less efficient) reducing reactions with different end-products. Instead of converting carbon to carbon dioxide, it is converted to methane (CH_4), sulphur compounds end as hydrogen sulphide (H_2S) and nitrogen compounds are reduced to release gaseous nitrogen.

The 'methanogenic' bacteria — those which produce methane — are very common. There are many species, all of them classified as belonging to the family Methanobacteriaceae. The 'marsh gas' that bubbles to the surface of stagnant water, especially in marshes, contains a variable, but usually high, proportion of methane.

Sometimes the methane (as well as phosphine [PH_3], which is also produced by bacteria) may ignite spontaneously and burn with a pale blue flame, visible at night. Such flames, known as 'ignis fatuus' ('foolish fire'), 'will-o'-the-wisp' or 'jack-o'-lantern', are said to occur over graveyards as well as

Below the surface, a marsh is airless and a place where anaerobic bacteria produce gases, such as hydrogen sulphide and methane (marsh gas) that enter the air.

marshes. It is hardly surprising that they have inspired so many legends. While they are certainly elusive, however, the flames are perfectly harmless — although the unburned gases are poisonous to humans, and people who wander from well-marked paths while crossing marshes or bogs may drown.

Bogs, marshes and graveyards occupy only a tiny part of the total land area of the Earth, but these are not the only places to provide the airless conditions in which the methanogens are able to thrive. Those which inhabit the digestive tracts of animals generate methane which is released as a by-product of digestion (flatus, which is more commonly known as 'gas' or 'wind'). Members of the Methanobacteriaceae also live in sewage sludge, permanently waterlogged soils, in the beds of many ponds and lakes, in tidal mudflats and in certain shallow, coastal waters. Gas which bubbles to the surface of Lake Kivu, Zaire, for example, is 25 per cent methane and the gas from Lake Beloye, in the USSR, is 80 per cent methane. Measurements of methane output, and of the quantity of methanogenic bacteria, have been made in the reduced sediments of the Gulf of California and the adjacent area of the Pacific, and in the Persian and Oman Gulfs and the adjacent part of the Arabian Sea. The bacteria were most common just below the surface of the mud, where they were found at up to 7100 cells per ounce of liquid silt (250 cells per gram), and in one sample more than twice that number. They were producing methane at a rate of 0.000048 to 0.0016 fluid ounces per ton of silt (0.0014-0.047 ml/t) per day. The quantity seems very small but this particular study, by Soviet scientists, concluded that, in places where the methane is trapped rather than escaping to the air, there is a real possibility of finding methane in marine sediments in quantities large enough to be exploited economically for industrial use.

Methane is also produced in the paddies where wet rice (as opposed to dry, or upland, rice which is grown in conventional fields, like other cereals) is cultivated. The enclosed paddies are flooded and the rice seedlings planted in shallow water. The mud in which they grow is airless, and provides a medium for the growth of methanogens, and the rice plants themselves release methane (for reasons that are not yet understood).

It is only under oxygen-free, reducing conditions that methane can be produced at all. This is because it is formed in a series of reactions that proceed in steps, at each of which the intermediary product is unstable and easily oxidized. Nor can the final product survive for long in the presence of oxygen. It breaks down — again in steps — to yield carbon dioxide and water. This is the fate of the methane that escapes into the air, and it explains why, despite this constant production of methane, the gas does not accumulate.

The atmospheric oxidation of (biologically produced) methane has two important consequences. The first is the augmentation of atmospheric water vapour. About 55 per cent of the water vapour in the air is a by-product of the oxidation of methane, and only 45 per cent results from the evaporation of surface water. (At very high altitude, this methane-derived water vapour is believed to contribute to the depletion of ozone over Antarctica; see page 160.)

Methane oxidation also removes oxygen from the air. It takes two molecules of oxygen to oxidize one molecule of methane to one molecule each of carbon dioxide and water. This helps to prevent any rise in the oxygen concentration, or 'partial pressure'. The partial pressure of one ingredient in a mixture of gases is the pressure (that is, the weight over a unit area, such as a square inch) which that gas would exert were all the other gases removed from the mixture. It is not the same as the proportion of the gas to the whole, because that is usually measured by volume rather than weight, and the weight of each gas is different.

The rate at which substances react with free oxygen is determined by the partial pressure of oxygen, rather than its atmospheric proportion by volume. Were the oxygen partial pressure to increase, by only a few per cent, exposed organic matter would be ignited by the smallest spark and the fires would be uncontrollable — until they had consumed enough oxygen to lower the partial pressure again.

Nitrogen's vital role
Nitrogen, comprising more than three-quarters of the atmosphere, reduces the oxygen partial pressure and the risk of uncontrollable conflagration. Were a significant amount of nitrogen to be removed from the air, the nitrogen-oxygen balance would be altered in favour of

oxygen — therefore, its partial pressure would increase. Because it is almost inert, nitrogen limits the spread of fire.

No one knows whether nitrogen was present in the primeval atmosphere or, if it was present, how much of it there was. Any that was there, however, would have been lost long ago had the Earth remained lifeless. The gas is inert, the two atoms that form its molecule being bound together strongly, with no opportunity for additional bonds to form with other molecules. Lightning, however, provides enough energy to break the molecular bond, separating the two atoms. In the presence of oxygen, free atoms of nitrogen are quickly oxidized, eventually to nitrate (NO_3), which is soluble in water and washed to the surface by rain. The thunderstorms that produce lightning occur only occasionally in any particular place but, over the world as a whole, it is estimated that about 1800 are happening at any one time — day and night, day after day, year after year — and each produces many lightning flashes. Over billions of years, therefore, all the nitrogen should have been removed from the air as soluble nitrate, and the oceans should be rich in it — a kind of 'nitrate soup'. The amount of nitrogen brought to the surface in this way varies widely from place to place, from 1.8 to 20 pounds per acre each year (2-22 kg/ha). At Rothamsted Experimental Station, in England, for example, 4 pounds per acre per year (4.5 kg/ha/yr) has been recorded.

The structure of every living cell is provided mainly by proteins. Enzymes, which promote chemical reactions necessary to life, are proteins, as are hormones, which move through plants and animals, each causing a particular organ to react in a certain way, and nucleic acids, such as DNA (deoxyribonucleic acid) and the RNAs (ribonucleic acids), by which genetic information is transmitted from parent to offspring and translated into the synthesis of more proteins. Proteins are made from very large, complex molecules, with thousands or even millions of atoms, arranged as chains of amino acids linked together by 'peptide bonds' (-CO-NH-). Nitrogen occurs in the bond, and amino acids themselves consist of an amino group (NH_2) joined to a carboxyl group (COOH), together with a little sulphur. The molecule can be arranged in many ways, so there are many amino acids

and, because the many amino acids can also be arranged in many ways, an immense number of different proteins. On analysis, proteins contain about 50 per cent carbon, 25 per cent oxygen, 15 per cent nitrogen, 7 per cent hydrogen, and a little sulphur. Nitrogen is essential to the synthesis of proteins and, therefore, to life itself, and nitrate is the form in which most nitrogen enters plants and, from them, animals.

It does not remain in the form of nitrate. Before it can be used to build proteins, the nitrate must be reduced to ammonia (NH_3) — and then used immediately, because ammonia is toxic and cannot be stored. When cells die, their decomposition releases the nitrogen again, as ammonia or ammonium (NH_4) compounds. Nitrogen is available, therefore, as nitrate, 'fixed' by lightning and dissolved in rain, and as ammonia or ammonium from the decomposition of dead organic material. Recycling is efficient — about 80 per cent of the nitrogen is recycled in a temperate forest — but lightning alone cannot supply nearly enough either to account for the stock which is recycled, or to make up the 20 per cent which is lost. In a single year, for example, a Scots pine forest takes up 40 pounds of nitrogen from every acre of soil (45 kg/ha), and a beech forest takes up even more — 45 pounds per acre (50 kg/ha).

Clearly, there is another mechanism by which inert nitrogen is being made available, and the agents involved are several groups of bacteria. In soils that are well aerated and not acidic (a pH of 6.0 is about the lower limit), there are bacteria belonging to the family Azotobacteraceae (*Azotobacter* is the best-known genus). They are able to utilize atmospheric nitrogen directly, incorporating it in their cells. The bacteria can live only in the presence of oxygen, and they feed on decomposing organic material. Where the soil is more acid, and poorly aerated, another group performs a similar function. These are anaerobic (they cannot tolerate oxygen), and belong to the family Bacillaceae (*Clostridium* is the best-known genus). Members of one or other family are found, usually in abundance, in almost every soil.

These bacteria live independently, but another group, of which *Rhizobium leguminosarum* is the best-known representative, establishes an intimate relationship with particular plants — the legumes. The plant root secretes a substance which attracts the bacteria and, when they make contact with a root hair, it curls around them. The bacteria then penetrate to the centre of the root, where they multiply to form a 'bacteroid', or 'nodule'. After that, the plant and its bacteria feed one another. The plant supplies carbohydrates to the bacteria, and the bacteria convert atmospheric nitrogen into compounds which can be used by the plant. The plants which form this relationship comprise the botanical family Leguminosae — the third largest of all families of flowering plants — and include many that are familiar and economically important, such as peas, beans and pulses of all kinds, and lupins. If you dig up a leguminous plant, very carefully so as to avoid damaging the roots, then gently shake off some of the soil, you will see the nodules of bacteria as pale 'lumps', somewhat larger than pinheads. When legumes are grown as farm crops, in a single year their bacteria fix between 50 and 200 pounds of nitrogen per acre (56-224 kg/ha).

When the plant dies, the bacteria die with it and the nitrogen is available for recycling by other plants. There are other soil bacteria which can live either independently, but unable to fix nitrogen, or in association with plants, when they do fix nitrogen. The plants which benefit from such relationships include alder, sea buckthorn and bog myrtle. Some cyanobacteria (of the genera *Anabaena* and *Nostoc*) can also fix their own nitrogen.

There is no difficulty, therefore, in explaining how nitrogen is moved from the atmosphere to the soil, and then how it moves from organism to organism as it is constantly recycled. There is a further set of reactions, however, involving yet another group of bacteria, which returns to the air about 10 to 20 per cent of the nitrogen in the soil — and sometimes much more than that in soils that are heavily fertilized.

Denitrification

This process, called 'denitrification', takes place only in the absence of oxygen, although the bacteria responsible for it are found in the air, as well as in soils and water. The bacteria obtain energy by reducing nitrate, in a reaction with two (or sometimes more) stages. In the first stage, nitrate (NO_3) reacts with hydrogen (H_2) to yield nitrogen dioxide (NO_2) and water

(H_2O), and in the second stage nitrogen dioxide reacts with hydrogen to yield water and gaseous nitrogen. If the soil is especially rich in nitrate, the bacteria may produce nitrous oxide (N_2O). Some of this also escapes into the air, and some remains in the soil and is reduced to nitrogen later, when the nitrate level falls.

Denitrification occurs in every soil, because in every soil there are regions from which air is excluded. Should soil become waterlogged, the rate of denitrification usually increases until it exceeds that of nitrogen fixation, so there is a net loss of nitrogen.

The principal denitrifying bacteria belong to the genera *Pseudomonas*, *Achromobacter* and, to a lesser extent, *Denitrobacillus*, all of which feed on organic matter (they are 'heterotrophs'). There are also a few species which can synthesize organic compounds from inorganic ones ('autotrophs'), denitrifying nitrate at the same time. *Thiobacillus denitrificans* obtains energy by oxidizing sulphur, for example, and *Micrococcus denitrificans* oxidizes hydrogen.

The fixation of atmospheric nitrogen is balanced by denitrification, which completes the nitrogen cycle. Over the last half century, the large increase in the use of nitrogen-based fertilizers has caused some distortion to the cycle, but mainly by accelerating it through the industrial fixation of nitrogen, which is then released into the environment. The cycle remains in balance, but it does not explain the presence of so much nitrogen in the air, mixed with oxygen.

The principal gas in the primeval atmosphere was carbon dioxide. If nitrogen was present, which is uncertain, it cannot have been more than a minor constituent and, possibly, it existed as no more than a trace gas — like argon in the modern atmosphere. Something must have released the very large amount that is present today. It seems unlikely that volcanic eruptions can have contributed so much nitrogen. The mixture of gases emitted by volcanoes nowadays contains no more than 5 or 6 per cent of nitrogen, and there is no reason to suppose that their chemistry was radically different in the early history of the planet. The more probable explanation is that the nitrogen was produced by denitrifying bacteria, which lived in the oceans and exploited the presumably abundant supplies of ammonia, and ammonium compounds. If denitrification began long before photosynthesis became established, atmospheric nitrogen could have accumulated to its present level before free oxygen was available to remove it. This would have increased the atmospheric pressure considerably. After that, nitrogen fixation and denitrification merely maintained the balance.

Several minor atmospheric constituents, which are also of biological importance, are moved through cycles by living organisms. Iodine, for example, is essential to animals, in very small doses. A lack of it causes goitre in humans and inhibits maturation in amphibians (supply iodine to an axolotl, which ordinarily spends its entire life as a larva, and it will develop into an adult salamander; withhold it from tadpoles and they will remain as tadpoles). Iodine is concentrated by certain seaweeds, in particular kelp (mainly *Laminaria* species),

which grows in huge beds in shallow coastal waters. The seaweeds produce methyl iodide (CH_3I), perhaps as an antibiotic, because it is extremely poisonous, and release it into the air, which carries it back over land, so sustaining an iodine cycle.

The sulphur cycle also involves many species of marine algae, but especially those of the coccolithophorids. The algae produce dimethyl sulphide, or DMS [$(CH_3)_2S$], some of which is broken down in the water by bacteria, but some of which escapes into the air, where it is oxidized, eventually to sulphate, which exists as crystals on to which water vapour condenses. The release of dimethyl sulphide increases cloud formation — and, over the oceans, far from the land, it is the principal source of the condensation nuclei that are necessary for cloud formation. The clouds carry the sulphur back over the land as sulphuric acid. This is a significant cause of 'acid rain' (see page 157), but it is also the principal means by which sulphur is returned from the sea to

At low tide on a sheltered seashore, large 'forests' of kelp (*Laminaria*) seaweed may be exposed. They make an important contribution to the transfer of iodine and other elements from the water to the air.

land-dwelling organisms — some is returned through the oxidation of hydrogen sulphide (H_2S) gas, produced by bacteria living in oxygen-free mud.

Living organisms influence the composition of the atmosphere to such a large extent that some scientists suggest it makes sense to regard the air as an entirely biological product. The organisms also maintain its composition, which has changed little for millions of years — apart from fluctuations in the amount of carbon dioxide, which is also due to biological activity. This idea, of the biological 'management' of the atmosphere (and oceans, for that matter) is the basis for the 'Gaia' concept.

This idea began to be formulated in the 1960s, when James Lovelock, its principal author, worked at the Jet Propulsion Laboratory, in Pasadena, California, as a member of the NASA team planning the exploration of the solar system. Its initial purpose was to provide a method for distinguishing a planet that supports life from one that is lifeless. 'Gaia' is now famous, and much that is non-scientific has been read into it, but while Lovelock and others have done much to refine it, its core remains unchanged. The concept holds that on a planet where life is found, the atmosphere will have been transformed, making it very different from the atmosphere of a lifeless planet — and that is how one may be distinguished from the other.

Voyages in space

On 20 July 1976, the spacecraft *Viking 1* landed on the surface of Mars. Its companion, *Viking 2*, landed some distance away on 3 September. Both *Vikings* transmitted observations made by their instruments and it was not long before we saw, and wondered at, the first pictures to be sent to Earth from the martian surface. They showed a dusty, rock-strewn landscape beneath a pale-blue sky. This was the kind of sky to which we are accustomed, and which we might have expected to see, but the colour was wrong. The first pictures were quickly withdrawn and re-issued with the colour corrected. The martian sky is not pale blue, it is pale pink (see page 51) — and no less beautiful for its strangeness.

We have no comparable colour pictures taken on the surface of Venus, although black-and-white photographs show a rocky landscape, but we should not expect blue skies over Venus, either. Its sky is more likely to be a pale brownish yellow.

There are clouds in the skies over Mars, but they are small and thin, and it never rains. Except when seasonal dust storms reduce the visibility, the night sky is brilliant with stars. No stars can ever be seen from the surface of Venus, however, for they are permanently hidden by the haze that colours the sky.

Terrestrial skies are blue because our air contains oxygen and oxygen is blue. The oxygen is the by-product of photosynthesis, a biological phenomenon which does not occur on Mars or Venus, where oxygen may be abundant but only as a constituent of compounds. There is little or no atmospheric oxygen on either of those planets.

The outer planets, beyond the asteroid belt, are too remote and different from Earth to make comparisons useful. Jupiter, 485 million miles (780 million km) from the Sun, is more than 300 times more massive than Earth and is made mainly from hydrogen and helium. Saturn, 870 million miles (1.4 billion km) from the Sun and 95 times more massive than Earth, is made partly from solid (metallic) hydrogen. Uranus, 1.9 million miles (3 million km) from the Sun and 14.5 times more massive than Earth, is made from water and molten rock enveloped in an atmosphere of hydrogen, helium and methane. It is similar to Neptune, about 2.8 billion miles (4.5 billion km) from the Sun and 17.2 times more massive than Earth. Pluto, the outermost planet at about 3.7 billion miles (6 billion km) from the Sun, is made from rock and ice.

Atmosphere of Earth and other planets

The extent to which biological activity has modified the composition of our atmosphere becomes evident when you compare Earth with its neighbours in the inner solar system. Mercury, the innermost planet, moves in a highly eccentric orbit, so its distance from the Sun varies more than that of other planets (except Pluto), but it averages about 36 million miles (58 million km), and the average daytime surface temperature is almost 800 °F (425 °C). Mercury is often described as having no atmosphere. This is not literally true. The planet has an atmosphere, consisting of carbon dioxide, but it is so tenuous — the surface pressure is about one-hundred-thousandth of that on Earth — that it cannot be compared with those of the other inner planets — Venus, Earth and Mars.

These three planets formed at the same time, in the same region of space, from the same stock of matter and in many respects they are similar. They orbit at different distances from the Sun, of course, so we should expect some variations in climate. Venus orbits at about 67 million miles (108 million km) from the Sun, Earth at about 93 million miles (150 million km) and Mars at about 142 million miles (228 million km).

All three have atmospheres and, in each case, the atmosphere was produced by volcanic outgassing. It is possible that small differences in the composition of the rocks on one planet, as compared to another, might have caused volcanoes to release their gases in different proportions, which would have produced differences in the compositions of the three atmospheres, but such variations are likely to have been slight.

There are other differences. Venus is only slightly smaller than Earth, but rotates on its own axis very slowly, giving it a day about 242.5 times longer than a day on Earth — and it rotates in the opposite direction to the Earth and Mars, so that on Venus the Sun rises in the west and sets in the east. Mars is much smaller than Earth, but its day is only 19 minutes shorter than our own. Because of the differences in distance from the Sun, a year lasts 247 terrestrial days on Venus, 365 on Earth and 687 on Mars. Earth and Mars are tilted on their axes, both by about the same amount, but Venus is almost upright. These astronomical differences are enough to ensure dissimilarities in the climates of the three planets, but not in the chemical composition of their atmospheres — and while the atmospheres of Venus and Mars are almost identical chemically, both are quite unlike the atmosphere of Earth.

The lower atmosphere of Venus consists of about 96.5 per cent carbon dioxide, about 3.5 per cent nitrogen, and about 0.007 per cent argon. There is less than 2 per cent water vapour and the merest trace of oxygen. The contrast with our own atmosphere — approximately 78 per cent nitrogen, 21 per cent oxygen and 0.03 per cent carbon dioxide — is startling, but becomes less so if you add to the atmospheric carbon dioxide content all the carbon dioxide that is stored in carbonate rocks, dissolved in the oceans and held in living tissues. This would give the Earth an atmosphere which is 98 per cent carbon dioxide and would remove all of its oxygen. The addition of carbon would also increase the total mass of the atmosphere but, even allowing for this, the atmosphere of Venus is much more massive than that of Earth, and the surface pressure is about 90 times greater than the terrestrial sea-level pressure. The average surface temperature on Venus is 891 °F (477 °C).

It is reasonable to assume that the early history of Venus was much like that of Earth, and that the volcanoes on both planets emitted similar mixtures of gases. If this is true, then the volcanoes of Venus must have released — and may continue to release, for Venus seems to be volcanically

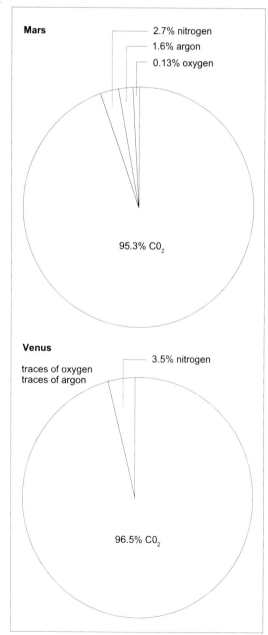

The atmospheres of Venus and Mars by volume

Saturn, photographed by the *Voyager I* spacecraft, has an atmosphere very different from that of Earth. The bright spots in the northern hemisphere may be huge convective storms, like our thunderstorms, but on a much larger scale.

active — substantial amounts of water vapour. No one knows how the atmosphere evolved, but it is possible to speculate. Suppose, for example, that the water vapour condensed, so that Venus had oceans. Carbon dioxide would have dissolved into the liquid water, leading to the formation of carbonates. This would have lowered the atmospheric pressure, and it would have been reduced further by the loss of water from the top of the atmosphere. There, sunlight would have provided sufficient energy to break down water molecules into hydrogen and hydroxyl (OH) ions, and the hydrogen would have been lost into space. At the same time, water would have been lost as its oxygen oxidized surface minerals.

After about a billion years, Venus would have lost most of its water. The formation of carbonates would have ceased and carbon dioxide, released from volcanoes, would have accumulated. This would have trapped long-wave radiation from the planetary surface, initiating a 'greenhouse effect' and the temperature would have risen, eventually to a level at which carbonate rocks began to release their carbon dioxide — just as on Earth the heating (kilning) of limestone to produce calcium oxide (quicklime) releases carbon dioxide. The atmosphere would have become more massive, dominated by carbon dioxide, and the greenhouse effect would have accelerated — as a 'runaway greenhouse effect' — until the system stabilized to produce the hot, dry conditions that exist now.

Venus is perpetually and totally blanketed by what appears from a distance as cloud (see page 51), but is really a thin, high-level haze with a base at an altitude of about 30 miles (48 km) and extending to a great height. At the surface, the horizontal visibility is about equal to that on an overcast day on Earth. In fact, it is possible to see much further than on Earth because, very close to the surface, the atmosphere refracts light strongly, parallel to the surface, so the horizon appears to be raised and on level ground an observer can see for perhaps hundreds of miles.

The surface is always hidden from outside observers, however, and they see the planet as a brilliant, white, almost feature-less object. The realization that Venus, in so many ways our twin planet, is covered by cloud led to intense speculation not so much about the possibility of life there as to the form such life might take. It was assumed, naturally enough, that the clouds were made from water droplets, and this led some people to suppose that Venus was largely covered by oceans and swamps supporting plants and animals of shapes, sizes and ways of life that were limited only by the imaginations of those who attempted to describe them.

The truth is more prosaic. Venus is one of the most arid places in the solar system, and nothing could possibly live there. Apart from the lack of water, no protein could survive the surface temperature — the peptide bonds that link the amino acids which form proteins are broken at much lower temperatures. Indeed, with its melting point of 621.5 °F (327.5 °C), lead would be a liquid on the surface of Venus. The cloud, or haze, contains a little water but the major constituent, accounting for three-quarters of its mass, is concentrated sulphuric acid.

The acid forms by reactions among the gases released by volcanoes. These include hydrogen sulphide (H_2S), carbonyl sulphide (OCS), sulphur dioxide (SO_2), carbon dioxide (CO_2), carbon monoxide (CO), and water vapour (H_2O). Inside the cloud, sulphur dioxide is oxidized to sulphuric acid (H_2SO_4), the reactions being catalysed by compounds of chlorine and hydrogen formed by the action of sunlight on hydro-chloric acid (HCl). The acid sinks below the cloud base and evaporates, becoming sulphur trioxide (SO_3). This reacts with carbon monoxide to form carbon dioxide and sulphur dioxide. At the same time, some of the carbonyl sulphide and hydrogen sulphide form pure sulphur and sulphur dioxide. At a higher level and just above the cloud top, the energy of sunlight causes the oxidation of the remaining carbonyl sul-phide and hydrogen sulphide, forming sulphuric acid and pure sulphur. They, too,

sink into the lower atmosphere where they are reduced by carbon monoxide and hydrogen, forming hydrogen sulphide and carbonyl sulphide once more.

These sulphur cycles, in which sulphuric acid is constantly being formed and then broken down again, are extremely stable. The atmosphere of Venus is in the state of chemical equilibrium it reached, probably billions of years ago, when the phase of runaway greenhouse warming ended. Even so, the clouds have some ameliorating effect. They absorb about two-thirds of the incoming solar radiation, so if they were absent Venus would be even hotter (on Earth, about two-thirds of the incoming solar radiation reaches the surface).

It is quite impossible for anything to live on Venus. Mars, on the other hand, is more promising, at least potentially, although at present the planet is lifeless.

Chemically, the martian atmosphere is much like that of Venus. It is 95.3 per cent carbon dioxide, 2.7 per cent nitrogen, 1.6 per cent argon, and 0.13 per cent oxygen. The atmosphere is very thin, however, and the surface pressure is a little more than 6 millibars (compared to 1000 millibars on Earth and 90,000 millibars on Venus). The average surface temperature is about -63 °F (-53 °C), making Mars a very cold place, but Mars does have seasonal weather, because of the tilt of its axis — like Earth but unlike Venus — and the average temperature covers wide variations. In winter, the temperature at the poles is around -190 °F (-123 °C), but at the Equa-tor the temperature can reach a comfortable 50 °F (10 °C).

In winter, the polar temperature is so low that atmospheric carbon dioxide solidifies and the icecaps, made from water ice overlaid by carbon dioxide 'dry ice', expand. In spring, as the ground warms, the carbon dioxide evaporates (technically, a substance that changes directly from the solid to the gaseous phase, without going through an intermediate liquid phase, is said to 'sub-lime') around the edges of the icecaps, as they retreat. This chills the adjacent air, because the latent heat of sublimation must be taken from the surroundings, and creates a large temperature difference between air close to the edge of the ice and air a little further away. The temperature difference generates strong winds, which pick up sand and dust, producing vast dust storms.

There are features on the surface of Mars which closely resemble the kind of dry river beds that are found on Earth, especially in deserts where rivers flow after the occasional rains, but then disappear. The only plausible explanation for the martian features is that once there were rivers on Mars. Water is still present, but only as thin ice beneath the polar icecaps, as permafrost and, possibly, as reservoirs of ice beneath the ground. Liquid water cannot exist at the surface because the atmospheric pressure is so low that the melting point and boiling point coincide and water, like carbon dioxide, changes directly between its solid and vapour phases.

If there is water on Mars, and if rivers once flowed, there must have been a time when conditions on the planet were very different from those which exist now. The atmosphere must have been much denser, for liquid water to exist, and the climate must have been much warmer. No one knows what happened early in martian history, but again we can speculate. Suppose that Mars began much as Earth began, with an atmosphere composed of volcanic gases that included large amounts of water vapour and carbon dioxide. These are both greenhouse gases and if enough of them were present — enough to produce a surface pressure of about 1000 millibars — the resulting greenhouse effect would have held the average temperature above freezing and the pressure would have inhibited evaporation. There would have been liquid water flowing in rivers, and probably lakes and seas as well.

Carbon dioxide from the air would have dissolved in the water, where it would have reacted with minerals such as calcium to form insoluble carbonates. Carbon dioxide is also believed to adhere readily to particles in the fine-grained 'soil', and substantial amounts may be held there. Little by little, this would have removed carbon dioxide from the air.

On Earth, some of the carbon dioxide lost from the air is replenished by volcanoes, but not on Mars. It is not that Mars has no volcanoes — indeed, Mount Olympus, some 14 miles (23 km) high, is the tallest mountain so far discovered anywhere in the solar system, and it was once a volcano — but that Mars is much less volcanically active than Earth. On Earth, volcanic activity is associated with the movements of the plates of which the crust is composed. On Mars, perhaps because the planet is much smaller than Earth, the crust is believed to consist of only one plate, so there can be no crustal movements to trigger volcanism.

As the carbon dioxide was removed, the greenhouse warming effect was reduced and the temperature fell, eventually to below the freezing point of water. At the same time, the atmospheric pressure also fell. Perhaps, within less than a billion years, the rivers had run dry and the seas turned to ice and been buried beneath the sand.

If you imagine a ball balanced at the top of an incline, the ball is said to be 'metastable'. It will remain where it is — it is stable — until something happens to give it a gentle nudge. Then it will roll down the incline and settle at the bottom. Once at the bottom, it will be truly stable and much more difficult to perturb. On Venus, the atmospheric conditions are very stable and it is difficult to imagine anything that might perturb them. They are like the ball at the bottom of the incline. Mars is not like that, despite its inhospitable appearance. Its atmosphere is thought to be metastable.

Suppose something were to happen — an increase in the amount of energy from the Sun, or some intervention by human explorers, for example — that initiated a slight warming. Carbon dioxide would sublime from the surface. This would produce a greenhouse effect, accelerating the warming and, if the temperature rose high enough to remove both polar icecaps, the overall reflectiveness of the planet's surface (its albedo, see page 148) would be reduced so it would absorb more heat and increase the warming still more. Water might then begin to evaporate from the permafrost, adding to the greenhouse effect, and the dust storms generated at the boundaries of the icecaps would also warm the air, because dust and sand particles absorb radiation. The atmospheric pressure would increase, perhaps carbon dioxide and water trapped in the soil would vaporize, and in time the rivers might flow once more.

Events such as these may have occurred more than once during martian history, but the climate they produced was even less stable than the one they replaced. In time, the carbon dioxide was lost from the atmosphere, the water froze again, and conditions reverted to their former state.

As the possibility of human exploration and even colonization of Mars draws closer, NASA (the United States National Aeronautical and Space Administration) is considering ways of achieving a more permanent 'terraforming' of Mars. This might involve injecting greenhouse gases into the atmosphere to trigger a general warming, and trying to establish dark-coloured lichens, which grow in the dry valleys of Antarctica, to lower the planetary albedo and sustain the warming. If it works, and theoretical studies suggest it should, the day may come when, in equatorial regions at least, humans can walk outdoors on Mars wearing breathing apparatus (the air would still contain very little oxygen), but no other specialized clothing. Life would have migrated to the dead planet, transforming it as its presence transformed the Earth.

Jupiter is 318 times bigger than Earth and, like the smaller Saturn, consists almost entirely of hydrogen and helium, with a turbulent atmosphere with fierce winds.

STRUCTURE OF THE ATMOSPHERE

We often describe the atmosphere as a kind of blanket enveloping the Earth. In some respects the image is misleading, however, because a blanket has a thickness you could measure, and this suggests that the atmosphere should also have a definite thickness. For many years this was the conventional view, modified a little by the realization that at the top the atmosphere does not end quite abruptly, but rather fades away through a region of increasing tenuousness. When plans were made to launch vehicles into space, no one doubted that the 'space' they were to enter lay beyond the uppermost reaches of the atmosphere, in what was essentially a vacuum.

Part of this image has survived, but the information acquired through space research programmes has necessitated a substantial revision of most of it. In particular, it has been revealed that the 'vacuum of space' is by no means empty, and our atmosphere does not simply end, so that if you were launched from the surface in a space vehicle you could say that at one moment you were inside the atmosphere and at a later moment you were outside it. The distinction is not so clear, and to ask someone the thickness of the atmosphere is like asking the length of a piece of string. Indeed, the concept of a 'vacuum' is now considered to be of only theoretical importance.

The space shuttles, which orbit the Earth at altitudes between 155 and 186 miles (250-300 km), are clearly in space — but are they outside the atmosphere? On the third shuttle flight, in March, 1982, the astronauts were asked to take photographs of an experiment installed in the payload bay. Their pictures, taken from the forward part of the shuttle looking towards the rear, showed the experiment, but they also showed a faint orange glow surrounding the tail fin and engine pods. Engineers on the ground also found changes in some of the materials used on the skin of the shuttle. The glow and these changes were caused — in different ways — by the atoms of oxygen and nitrogen through which the shuttle had flown. High though it was, the craft had not quite left the atmosphere. The 'air' was extremely thin — about 16 billion atoms in every cubic inch(1 billion per cu cm)

compared with about 500 billion billion per cubic inch (3 x 10^{19}per cu cm) at sea level — but in no sense could it be called a vacuum. There is enough air, even at that altitude, to exert a slight but measurable drag on spacecraft.

All gases, including air, are easily compressed. A planet retains its atmosphere because of the gravitational force it exerts, but this force decreases rapidly with increasing distance from its source. The gradual 'fading out' of the upper atmosphere is due to the combination of these two factors.

The air that we breathe, through which we move and without which we cannot survive is a material substance. We cannot see it, except as the blueness of the sky, but physically it is as real as any of the solid objects we see around us — as real as a house or a tree. Being a real substance, air has mass and it also has weight.

Mass and weight are not the same thing, although we commonly use the words as though they were synonymous. The mass of an object is a measure of its inertia — of the force needed to move it if it is at rest. If two bodies of equal mass collide — picture two balls on a snooker or pool table — they will be affected equally by the collision. If the balls are of different masses, the one with the smaller mass will move faster and further than the one with the greater mass. At the same time, mass is a measure of the gravitational attraction a body exerts on other bodies within its range.

Gravity is a very weak force that becomes significant only when a body has a very large mass, but you can demonstrate it for yourself with two large, steel ball bearings. Place them on a very smooth, absolutely level surface and then move them closer together, in steps, until they are so close that they roll towards one another by themselves and you hear the click as they collide. It is their mutual gravitational attraction that draws them together.

The amount of force needed to move a body from a state of rest, and the gravitational attraction a body exerts on other bodies, are two ways of describing the same thing — a property the body possesses and that is inherent to it. The mass of a body is constant regardless of its location in the universe. Scientifically, mass is measured in kilograms. If your body has a mass of, say, 70 kilograms (equal to 154 pounds), travelling into space will not affect it.

Its weight, on the other hand, will not remain constant. Were you to visit Mars, for example, you would weigh about 27 kilograms (58.5 lb) although your body would still have a mass of 70 kilograms (154 lb). Weight is the force of gravitational attraction acting on a mass. Because it is a force, scientifically it is measured in newtons (1 newton, for which the symbol is N, is the force required to give a mass of 1 kilogram an acceleration of 1 metre per second per second). Weight is not a physical property possessed by a body. It depends on the gravitational force to which the mass of the body is subjected, and that can vary.

On the surface of the Earth, such a subtle distinction is not important. It was made by scientists who, quite sensibly, gave a value of 1 to the average gravitational force at the Earth's surface ('average' because it is not precisely the same everywhere). This means that for bodies at the surface of our planet, mass and weight are equal. The distinction becomes important, however, when you consider bodies that are not at the surface of our planet, because the strength of the gravitational force is equal to the combined masses of the bodies affected by it, but also inversely proportional to the square of the distance between them. This phenomenon was first described by Isaac Newton, the English scientist (1642-1727) and is known as his 'law of gravitation' or the 'inverse square law'.

The Moon, for example, may once have had an atmosphere of some kind, but it was unable to retain it because, although that atmosphere may have had a considerable mass, it had very little weight because the mass of the Moon is too small to exert a gravitational force strong enough to have given it that weight. Near the top of its atmosphere (if it had one) lunar gravity would be reduced according to the inverse square law, and the gravitational force exerted by the much more massive Earth would strip away the gases, despite the even greater distance. Today, all that remain are atoms and molecules of some heavier gases, including water vapour, that move freely just above the lunar surface. These constitute an atmosphere of sorts, but one so rarefied that it has no measurable effect.

The martian atmosphere also suffered losses. It may once have had 30 times more nitrogen than it has now. Two isotopes of nitrogen occur naturally (N-14 and N-15) and always in the same proportions. N-14, the lighter of the two, was lost because Mars was not massive enough to exert the gravitational force needed to retain it. Earth retained its N-14 and it is by comparing the proportions of the two isotopes in the atmospheres of the two planets that the nitrogen loss from the martian atmosphere can be calculated.

When atoms are lost from a planetary atmosphere they do not cease to exist, and neither do they escape from the solar system. They join the great, thin, swirling, invisible cloud which occupies the space through which the planets and their satellites move.

The Sun, too, has an atmosphere and, because the Sun accounts for more than nine-tenths of the total mass of the solar system, its atmosphere is much larger than that of any planet. The solar atmosphere extends far beyond the orbit of Earth, and at a height of about 50,000 miles (80,000 km), our atmosphere merges imperceptibly with it.

Comet tails provide clear evidence for this. Not all comets develop them but, for those which do, the tails consist of particles detached from the cometary surface. The particles reflect sunlight, which makes them visible, occasionally to the naked eye, and comet tails sometimes extend for thousands of miles from the head of the comet itself. The tail does not trail behind the comet, however, like the smoke from a steam locomotive. It always points directly away from the Sun, regardless of the direction in which the comet is moving. This is because the tail is being 'blown' by the 'solar wind', a stream of electrons and ionized hydrogen and helium atoms ejected from the Sun. In our region of space the solar wind consists of up to more than 160 particles in every cubic inch(10 particles per cu cm). It is a thin wind, but a wind for all that, and a fast one. It 'blows' at up to 1,800,000 miles per hour (800 km/s).

Layers of the atmosphere
At lower levels, the atmosphere forms layers which are distinct and separated by boundaries that tend to prevent the passage of molecules and solid particles in either

Thin and wispy, cirrus cloud consists of tiny ice crystals. It forms close to the tropopause and often marks its location.

The atmospheres of Mars and Venus consist mainly of carbon dioxide. (Opposite, top) That of Mars is very thin. The sky is pink because of dust particles. The climate is cold. (Opposite, bottom) Venus has a very dense atmosphere and the temperature at the surface is high enough to melt lead.

direction. These layers form a set of concentric spheres surrounding the Earth and each has a name ending with the suffix -sphere. This structure is controlled by temperature.

The lowest layer is called the troposphere. The Earth's surface provides its lower boundary and it extends upwards as far as the tropopause. The height of the tropopause varies from place to place and season to season — occasionally it can descend to surface level — but generally it is at about 10.5 miles (17 km) over the Equator and sometimes as low as 5 miles (8 km) over the poles, although usually it is at about 6 to 7.5 miles (10-12 km).

The really huge storm clouds that bring thunder and lightning are more common in some parts of the world than others — they are especially common in the tropics — but if ever you have a good view of one it will reveal the location of the tropopause. Such clouds are called 'cumulonimbus' (from the Latin *cumulus*, meaning 'heap' and *nimbus* meaning 'cloud') and they stand as great,

threatening towers. The top of a big cumulonimbus is often flattened with what looks like a tail trailing to one side, giving the cloud top the approximate shape of a blacksmith's anvil — and these clouds are often called 'anvil' clouds. The flat top of the 'anvil' marks the level of the tropopause, and demonstrates clearly that this is a real boundary which gas molecules and water droplets can rarely cross. If you wonder about the height of such a cloud, in middle latitudes the 'anvil' probably marks an altitude of a little more than 9 miles (15 km), and is higher than most civil aircraft are able to fly.

The existence of the tropopause is due, in the first place, to the fact that air in the lower part of our atmosphere is transparent to all of the solar radiation to which it is exposed (but not all the radiation to which the planet is exposed, because some radiation is deflected or absorbed at a much greater height). The radiation passes through the air without affecting it and

(Top) At high altitude, the air density is much reduced, the temperature is much lower than at the surface, and humans must wear warm clothing and carry oxygen. (Bottom) The troposphere, extending from the surface to the tropopause, is the layer of the atmosphere in which temperature decreases with height, water vapour evaporates and condenses and, therefore, weather occurs.

warms the land and sea surface. Contact with the warmed surface then warms the air.

When a substance is warmed, energy — as heat — is imparted to its molecules. If the substance is a solid, this enables the molecules to vibrate more vigorously. If it is a fluid (liquid or gas), the molecules already move independently of one another, and the addition of energy allows them to move faster. When fast-moving or vigorously vibrating molecules strike our skin, nerve endings detect their impact and our brains interpret the sensation as warmth. We commonly describe the temperature of substances in terms of our sensory response to them. This is perfectly satisfactory for ordinary purposes but, when considering what happens in the atmosphere, it is important to distinguish between our subjective sensation, of warm or cool air or water, and the actual cause of that sensation — the energy possessed by molecules. When molecules accelerate, the distances between them tend to increase — they try to fly away

from one another. This exerts a pressure — of 'pounding molecules' — on their surroundings. If those surroundings are rigid, the phenomenon can be harnessed to do useful work — it is the principle behind all engines (such as steam and internal combustion engines) that burn fuel to raise the temperature of a working fluid. If the surroundings can be compressed, however, the warmed fluid expands to occupy a greater volume and, therefore, it becomes less dense. This means that the mass of a unit volume (such as a cubic foot) of it decreases. The reduction in the mass of the warmed volume of fluid also decreases its weight — the force exerted by gravity on its mass. It is held by the planet a little less firmly than it was before it was warmed. In particular, it is held less firmly than the surrounding fluid, which is cooler. So it will rise, like a bubble of air through water.

Gases, including air, can be compressed quite readily. We compress air every time we pump up a tyre or a balloon — it is easy.

Cumulus and cumulonimbus clouds form by convection, as air, warmed by contact with the surface, rises rapidly, cools, and its water vapour condenses. (See also pages 60-1.)

The weight of the atmosphere also compresses the air of which it is composed. The great mass of the atmosphere is given weight by the gravitational attraction that draws it towards the centre of the Earth but, because the molecules in a gas are not attached to one another, the force acts on each of them individually. At the bottom of the atmosphere, the air molecules are closer together than they are at higher levels and, therefore, the density of the air decreases with increasing height. The troposphere contains about three-quarters of the entire mass of the atmosphere — but only a minor proportion of its total volume.

If you fly at an altitude of more than about 10,000 feet (3000 m), oxygen must be added to the air you breathe, and you may need extra oxygen at lower levels if you are exerting yourself strenuously. This is not because the composition of the air changes at this height — it still contains about 21 per cent oxygen — but because the air itself is less dense and a 'lungful' contains fewer molecules of all the atmospheric ingredients, including oxygen. People who live at high altitudes adapt to the low air density by producing more haemoglobin, so their blood compensates for the reduced amount of oxygen by making more efficient use of the oxygen available to them.

The 'parcel' of air that has been warmed and made less dense by its contact with the Earth's surface rises, but it rises into a region of steadily decreasing air density. As it does so it also cools. The energy with which its molecules move more rapidly becomes dissipated — 'used up' — and, as it rises away from the warm surface, it loses contact with its source of warming. Its molecules move more slowly and, to some extent, they move closer together again but, because the 'parcel' is now surrounded by less dense air in which the molecules are further apart than they would be close to the surface, the 'parcel' never returns to the size and density it had originally. Eventually it will be surrounded by air at a density equal to its own. When it reaches this level — the tropopause — it can rise no further because a unit volume of it weighs no more and no less than a similar volume of the surrounding air.

Composition of the layers
The tropopause is a cold region. If you have walked in high mountains, or seen pictures of mountaineers, you will know that you need warm clothing if you are to feel comfortable at high altitude. There are quite wide variations but, on average, the air temperature in the troposphere falls by about 3.6 °F for every 1000 feet of increasing altitude (6.5 °C per km). At the tropopause, the average temperature is -76 °F (-60 °C) but the average contains variations. In winter, for example, the temperature at the equatorial tropopause — the coldest region at all times of year because it is at the greatest altitude — averages about -112 °F (-80 °C). Immediately above the tropopause, the air temperature no longer decreases with increasing height and, in some places, there is a layer of slightly warmer air. This is called a 'temperature inversion', and it prevents the lower air from rising further because to do so the rising air would have to pass through a region of lower density.

There is a slight leakage in both directions across the tropopause but, in general, gas molecules and solid particles can cross only if they are contained in 'parcels' of air that are less dense than the air surrounding and above them. This rarely happens because they are far removed from the principal source of atmospheric heating — contact with the surface.

The atmospheric layer above the tropopause is known as the stratosphere, and it extends upwards to a height of about 30 miles (50 km) above the surface. Temperatures in the stratosphere change according to the seasons but, in the lower stratosphere, there is no change in temperature with increasing height. This is because the stratosphere is effectively isolated from the principal source of atmospheric heating — contact with the ground surface. Above about 12 miles (20 km), however, the situation changes and the temperature begins to increase with increasing height. This is especially marked in summer. The pattern is more complex in winter, when there is usually very cold air in the middle stratosphere over the middle latitudes. The relationship between temperature and altitude changes because, at this height, the atmosphere begins to absorb some of the incoming solar radiation.

Air is transparent to visible light, but oxygen absorbs and, therefore, is opaque to radiation in the ultraviolet (UV) part of the electromagnetic spectrum (see page 71). As radiation passes through the atmosphere, the chance that it will encounter an oxygen (or any other) atom or molecule increases the further it travels because the distance between particles decreases as the density of the air increases. At about 14 miles (22 km) UV is absorbed in significant amounts. This absorption produces a concentration of ozone, in the ozone layer (see page 159), and it also warms the air, because the process transfers energy to the absorbing particle, thus raising its temperature. At the top of the stratosphere, the absorption of UV raises the temperature to about 32 °F (0 °C), but the air is so rarefied — the pressure is about 1 millibar (mb), which is one-thousandth of that at sea level — that, although particles have energy, there are too few of them for much ozone to form.

The upper boundary of the stratosphere is called the 'stratopause', and the region above it is the mesosphere, extending to a height of about 50 miles (80 km). The air density continues to fall with increasing height, from about 1 mb at the stratopause to about 0.01 mb at the upper boundary of the mesosphere. It is also a region in which temperature once again decreases with increasing altitude, from about 32 °F (0 °C) at the stratopause to -130 °F (-90 °C) at 50 miles (80 km).

You might think that the mesopause, the upper boundary of the mesosphere, marked the top of the Earth's atmosphere, but 'meso' means 'middle', and above the mesopause there lies the thermosphere. This is a layer in which the absorption of solar radiation causes the temperature to rise with increasing height, perhaps to about 1700 °F (930 °C) at 220 miles (354 km). The gases at this height are so rarefied, however, that the air does not feel warm and orbiting satellites are not warmed by it. The ionosphere (see page 71) forms part of the thermosphere above about 62 miles (100 km).

At between about 300 and 470 miles (480-755 km) — there is no definite boundary — the thermosphere becomes the exosphere. This merges with the magnetosphere, where particles are dominated by the Earth's magnetic field, concentrated most densely between about 1860 and 10,000 miles (3000-16,000 km), and extending towards the Sun to about 35,000 miles (57,000 km), with a tail reaching much further on the side away from the

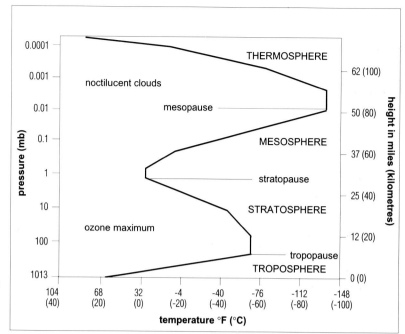

Sun. It is the magnetosphere which merges with the solar atmosphere.

Journey into space

A space traveller, departing from Earth for a destination elsewhere in the solar system, would pass through all the layers of our atmosphere. Unfortunately, such a passage would be far too brief to allow any examination or appreciation of the regions that were traversed. A more leisurely means of ascent is required.

We are surface-dwellers, and bound to the surface more firmly than we may suppose in these days of air travel. The air with which we are familiar is contained in the troposphere. Many of us have flown through tropospheric air, and some of us may have crossed the tropopause — barely — but very few people have ventured beyond the lowest regions of the stratosphere, and only astronauts have visited the outer regions of the atmosphere. A journey of a mere 100 miles (161 km) to the region directly above our heads is beyond us, and we know far less about what happens there than we do of events on the other side of the world, but at the surface.

If you were to climb slowly and steadily through the atmosphere, you would see everything, but feel nothing, if you were inside an aircraft, so we must suppose that you could remain exposed to your surroundings — in the basket below a balloon, perhaps. You would need to take special clothing and equipment, for protection

The layering of the atmosphere is related to the way air temperature changes with increasing height, with boundaries (the tropopause, stratopause and mesopause) occurring where the temperature remains constant over a short vertical distance, inhibiting the exchange of gases. Air pressure decreases at a fairly regular rate as height increases.

The wind is produced by differences in air pressure and the greater the difference the stronger the wind. Here, in the Mataamata Mountains, Tunisia, the wind carries light, powdery sand, causing a sandstorm.

when conditions became intolerable, and you would also need instruments, to measure changes your senses would fail to register.

Your first impression, on an average day, might be of a sharp increase in the wind speed as soon as you were well clear of the surface. The air is seldom still, but it is slowed by friction as it passes across the ground and encounters obstacles such as trees and buildings. You would feel this change at a height of 200 feet (60 m) or less, and you can often see it from ground level. When the wind is too light to disturb even the smallest twigs on the trees and scraps of paper fall almost vertically, low-level clouds are often moving across the sky quite

steadily and fairly rapidly.

A little higher, perhaps you will approach the base of cloud. As you enter the cloud you will need to don waterproofs, because clouds are composed of water droplets and will soak you very quickly. The cloud, feeling like fog, will seem cold. The reason, you may think, is quite straightforward. If you cool water vapour it will condense. This is why the windows of a car 'steam up' on a cold day — the passengers exhale water vapour which condenses on to the cold windows. This is broadly true, but in a cloud matters are not quite so simple. The water droplets are cold, but the air between them is much warmer.

The melting of ice and evaporation of water involve major physical changes — known technically as changes of 'phase'. Water is not simply warm ice and water vapour is not just warm water, but in all its phases water is the same substance, H_2O. What happens at a phase change is that the molecules acquire sufficient energy for them to move more freely. Ice consists of molecules locked together. If they can acquire enough energy, they will separate and move freely in relation to one another, and the solid will turn into liquid. If the molecules of the liquid acquire still more energy, they will fly off by themselves into the air, and the liquid will turn into a gas. At each

change, additional energy — as heat — is needed to give each molecule the necessary freedom to break free, to break the bounds that determine the distance between it and its neighbours. This is called 'latent' heat, because all it does is to change the phase. The temperature of the water — at the freezing or boiling point — is not altered at all by the absorption of latent heat. When the phase changes in the reverse direction — and water vapour condenses or water freezes — the same amount of latent heat is released. Different compounds absorb and release different amounts of latent heat. In the case of water, at freezing point, it takes 271,000 calories of heat to vaporize 1 pound (2,500,000 joules per kg), and 36,255 calories to melt 1 pound of ice (335,000 J/kg). In a cloud, the latent heat released by the condensation of water vapour is imparted to the surrounding air which is warmed. This also explains why it often feels warmer when snow falls after a period of intensely cold weather, and why the spring thaw produces a distinct chill.

The amount of water vapour air can hold depends on the air temperature. Not only can warm air hold more water vapour than cool air, but its capacity to do so increases with rising temperature at an accelerating rate. At 50 °F (10 °C), for example, a cubic foot of air can hold 0.009 ounces of water vapour (9.4 g/cu m). At 68 °F (20 °C), a cubic foot can hold 0.02 ounces (17.3 g/cu m), and at 86 °F (30 °C) it can hold 0.03 ounces (30.4 g/cu m). As the temperature increases from freezing to 50 °F (10 °C), the water vapour capacity of a cubic metre of air increases by 0.16 ounces (4.6 g), but an increase from 77 °F (25 °C) to 95 °F (35 °C) — in both cases a rise of 18 °F (10 °C) — increases the water vapour capacity by 0.58 ounces (16.55 g), about three-and-a-half times more.

The temperature at which a volume of air is fully saturated with water vapour is called its 'dew-point' temperature. If air is cooled still further, in most cases (but air can be supersaturated — see below) the vapour will start to condense. Obviously, the dew-point temperature depends on the water-vapour content of the air — the drier the air, the lower the dew-point temperature will be. The amount of water vapour in the air, expressed as a percentage of the amount needed to saturate the air at that temperature, is the 'relative humidity' of the air, and it reaches 100 per cent at the dew-point temperature. The relative humidity is the figure most commonly quoted in weather reports which give the humidity, but meteorologists also use other measures. The 'absolute' humidity is the actual mass of water vapour present in a given volume of air (usually 1 cu m), and the 'specific' humidity is the mass of water vapour present in a given mass of air (usually 1 kg).

It is possible for the dew-point temperature to be below freezing. In this case the vapour changes directly into ice without passing through the liquid phase, and the clouds are composed of ice crystals rather than water droplets.

As you climb, you start to feel cold because, of course, the temperature of the air decreases as you move higher. The rate of change is called the 'lapse rate', and its average value is about 3.6 °F per 1000 feet (6.5 °C/km). This is the change your on-board thermometer records. It is called the 'environmental lapse rate'. If a volume of air is made to rise through the surrounding air, however, it will expand and cool without exchanging heat with the air around it and, should it be forced to descend, it will warm in the same way. Such changes in temperature, that do not involve the surrounding air, are called 'adiabatic', and the actual rate depends on whether or not the air is saturated with moisture. In dry air, the (dry) adiabatic lapse rate is 5.4 °F per 1000 feet (9.8 °C/km), so if you began your ascent when the air temperature was, say, a comfortable 60 °F (15.5 °C), and you rose with a 'parcel' of air that was not mixing or exchanging heat with the air around it, by the time you reach 2000 feet (610 m) it will have fallen to about 50 °F (9.5 °C).

As you enter the base of your cold, wet cloud, you cross the boundary at which the temperature falls to the dew point and once you are inside the cloud, the lapse rate changes — to the 'saturated adiabatic lapse rate'. The change occurs because the latent heat of condensation, released as the cloud droplets form, warms the air between the droplets and this reduces the lapse rate. The extent to which it does so depends on the initial temperature of the air. In very warm air, the lapse rate may be halved, but at very low temperatures, around -40 °F (-40 °C), the two lapse rates are about the same. You may feel little benefit from this, for the difference is not great, but it can affect the

way your cloud forms. If it forms in very warm, rising air, the release of latent heat and the associated reduction in the lapse rate will mean the air inside the cloud is markedly warmer than the air outside, so it will continue to rise. As the air within it rises further, its falling temperature will cause more water vapour to condense, releasing more latent heat, and the cloud will grow higher and higher, as a 'heap' or 'cumulus' cloud.

The air inside a cumulus cloud is unstable. This means that, once it starts to rise, it will tend to keep rising. This condition occurs when the environmental lapse rate lies between the dry and saturated adiabatic

lapse rates — which it often does. As you rise into the cloud, you may feel yourself accelerating upwards. There are (possibly apocryphal) airmen's yarns of pilots parachuting into very unstable air and finding themselves travelling upwards. If the cloud forms in stable air, however, your vertical progress will be more dignified.

Eventually you will reach a level at which the dew-point temperature is lower than the actual air temperature and, therefore, water ceases to condense. You will notice this boundary, because it will mark the top of the cloud, and you will enter clear air. There may be more cloud above you, however, where condensation resumes. Rise

through that, and there may be yet more cloud at a still greater height, but this will be thin, wispy, and of the 'cirrus' type ('cirrus' is a Latin word meaning 'curl'). This cloud is made from very small ice crystals, and it is so thin that you will barely notice it as you pass through it.

The close relationship between the temperature of the air and the amount of water vapour it can hold means that the higher you climb the drier the air becomes because it constantly grows colder. In effect, the environmental lapse rate 'squeezes' water from the air. About half of all the water vapour in the atmosphere is held at below 6500 feet (1980 m).

When the temperature inside a cloud is below freezing the water condenses as ice crystals. If these link together they may become heavy enough to fall as snowflakes.

Light aircraft, like these Optica Aiguts, fly at low altitudes, seldom climbing above about 7000 feet (2135 m).

Although, over the planet as a whole, the air is very moist, water passes through the atmosphere quickly. On average, a water molecule which enters the air will remain there for about 10 days. By the end of that time it will have returned to the surface as rain or snow. The system is very dynamic indeed.

It is not only water that is 'squeezed' from the air by the rising temperature. So are small, solid particles — mainly of dust. Water vapour condenses when the air temperature falls to the dew-point, but only if a surface is present on to which it can condense. At ground level it will condense on to any surface — as it does on to car windows — but in the air it requires solid particles. In the absence of such particles — called 'cloud condensation nuclei' — the temperature can fall well below its dew-point. When this happens the relative humidity rises to more than 100 per cent, although it seldom exceeds 101 per cent, and the air is said to be 'supersaturated'. In extremely clean air, however, where there are no cloud condensation nuclei at all, a small drop of pure water may need a relative humidity of rather more than 100 per cent

to sustain it, and a very tiny droplet may need a relative humidity of more than 300 per cent. The difference between a small drop and a large one is due to the ratio between the volume and surface area of the droplet. A very small droplet has a small volume and a relatively large surface area. Unless the air is highly supersaturated, and, therefore, resists strongly the addition of more vapour, the droplet will evaporate again as quickly as it forms. A larger droplet, with a larger volume in relation to its surface area, evaporates less readily.

Not all particles are suitable as cloud condensation nuclei. Very large ones will fall under their own weight, and so they are unlikely to remain airborne long enough for water to condense on to them. Very small ones will not fall quickly but, because so little water can condense on to them, the air must be highly supersaturated for condensation to occur. The size, of particles and the droplets which form around them, is measured in micrometres, which used to be called 'microns' (a micron is one-millionth of a metre, or about 0.00004 in). A particle with a diameter of 10 micrometres (0.0004 in) or more will fall quite fast, and is too large to be a cloud condensation nucleus. One with a diameter of 0.001 micrometre (0.0000004 in) is too small. Between the two extremes, the suitable particles include dust, soil particles blown into the air by the wind, smoke, and sulphur dioxide and salt crystals. The last two are especially effective, because they are soluble and, once they have dissolved into the water condensing on to them, evaporation becomes more difficult.

The higher you climb, therefore, the cleaner the air becomes. You will be able to remove your smog mask as you drift across industrial cities, because, above the cloud tops, you will see, but be beyond the reach of, their emissions. When you reach 6000 feet (1830 m) — and under most conditions near the upper limit for industrial haze — you will be about as high as private light aircraft usually fly — and often they fly much lower.

At around 10,000 feet (3000 m), you should don your oxygen mask. As your analytical instruments will tell you, the composition of the air has not changed but there is less of it. The air pressure at this height is about 700 mb — only 70 per cent of the sea-level pressure. You will also need really warm clothing. If the temperature at ground level was 60 °F (15.5 °C), at this height it will be about 24 °F (4 °C). At around 20,000 feet (6100 m) you should increase your oxygen supply and put on your arctic clothing, because the temperature now will be around -12 °F (-24 °C) — about what you might expect on a January morning in Chicago. The air pressure is now about 500 mb — half its sea-level value.

You have travelled vertically no more than about 3.75 miles (6 km), but you might as well be in a different continent and a different season, yet the tropopause is still far above your head. As you rise past about 30,000 feet (9150 m), you leave virtually all the water vapour behind. You may see the last traces of it as thin cirrus clouds. The air is now 'bone' dry and almost completely clean. The pressure is about 240 millibars and the temperature is about -58 °F (-50 °C). Should your oxygen supply fail, you will die very quickly.

At this height, if you look upwards, you will see that the sky is a distinctly darker shade of blue than it appears from ground level. You are at about the altitude at which medium-haul jet airliners fly. Jet engines operate most efficiently (in terms of the fuel burned for the thrust delivered) in the upper troposphere but, on short journeys of up to about 200 miles (320 km), the fuel consumed in the climb, and the short time the aircraft can spend at its cruising height, combine to cancel out the advantage, so they fly lower.

Intercontinental airliners fly somewhat higher, closer to the tropopause which, in middle latitudes, is usually at a little over 40,000 feet (12.2 km). At this height, the air is still 78 per cent nitrogen and 21 per cent oxygen but it contains no more than an occasional molecule of water vapour and, except under very unusual circumstances, no solid particles. As you pass across the tropopause, you enter a region of very clean air indeed. You will not be able to breathe it of course. The air pressure is about 120 mb, and, in middle latitudes, the temperature hovers around a frigid -67 °F (-55 °C).

If you look at the ground, far below, as you drift up towards the tropopause, at around 40,000 feet (12.2 km), you may find yourself moving gently eastwards. The motion will appear slow because the ground is so far away — for the same reason that aircraft appear to move slowly when you are on the ground and they are very high. In fact, you are probably moving at more than 100 miles per hour (160 km/h). You have encountered a jet stream (see page 87) and you should keep your eyes open for westward-bound aircraft, using the jet stream as a tail wind — although you may not see them, even if they are there. Except when you look downwards, the sky is completely empty. There is nothing at all to see, and so there is nothing on to which your eyes can focus. You may peer around as vigilantly as you wish, but your eyes will be quite relaxed, focused about 3 feet (1 m) in front of you. A distant aircraft, no larger than a speck, will be invisible to you — but may be approaching very fast indeed. At this height, the clear sky provides limitless visibility — but only radar can tell you whether it is really as empty as it looks.

As you reach and cross the tropopause, you will leave the jet stream and enter a region of great tranquillity. You will drift across it effortlessly, but this boundary bars the passage of most gas molecules and particles. Should they cross, and some do, they will be trapped in the stratosphere for a long time. A particle will remain airborne in

the troposphere for a matter of days before it falls to the ground or is washed out by rain. A particle that is carried into the stratosphere may remain there for up to two years. This can happen. Extremely violent volcanic explosions sometimes eject a mixture of gases and dust at such high temperatures that the cloud rises rapidly through the atmosphere, crosses the tropopause and, by the time its ascent is checked, it is in the stratosphere, and held there. In the past, atmospheric tests of nuclear weapons also injected material into the stratosphere, and some gaseous pollutants may do so if they are sufficiently stable chemically to survive for the time it takes for them to drift slowly and gently upwards and across. The long residence time of stratospheric materials means that any effect they may have — on the climate, for example, or the ozone layer — is prolonged. It is why scientists are especially concerned about air pollutants that cross the tropopause.

For some time now, common sense should have warned you that to avoid frostbite you must not expose any part of your skin. Your layers of clothing will protect you against the cold and, for a while yet, they will protect you against the radiation to which you are being exposed. This increases steadily as you climb higher. Cosmic radiation (see page 143) is not yet a serious problem — although it might become one were you to spend a long time at this very high altitude — but you are entering a region where ultraviolet (UV) radiation is more intense. You will have some difficulty taking refreshment, partly because you must not remove your oxygen mask, and partly because the air pressure is now so low that any liquid will evaporate instantly if it is exposed to the air. From now on, your meals will have to consist of liquids held in pressurized containers, and you will have to drink them through a tube in your face mask.

The air will become no colder as you continue to rise, but it will become more rarefied. At 60,000 feet (18.3 km), you will be at about the cruising altitude of Concorde and many modern military aircraft. You may be struck by the almost complete silence and, like most people who fly alone at such a height, probably you will experience an intense and disorienting loneliness. The region you have invaded, no more than 11 miles (18 km) from the noise and bustle

of home, is remote, unpopulated, and the people so far below are quite unaware of your existence. You are more alone than you could ever have imagined — and there is still a long way to go.

When you reach 70,000 feet (21.3 km) your instruments may inform you of certain changes. The air is just a little warmer — though only a thermometer will detect the difference — and its composition has changed. It contains a small amount of ozone. There is not much of it — between 1 and 10 parts of ozone to a million parts of other gases — but this is the region of highest ozone concentration. You are in the 'ozone layer', and the slight rise in temperature is caused by the absorption of UV radiation. The air pressure is now about 44 mb and the sky above you is quite dark. It might be worth watching out for your first glimpse of a star. Soon they will be clearly visible, even in daytime.

A few military reconnaissance aircraft fly still higher, at up to 90,000 feet (27.5 km), but the altitude record for an aircraft is higher still. In 1961, a Soviet pilot climbed to 113,892 feet (34.7 km) in a jet-powered aircraft, and in 1963 an American pilot reached 354,200 feet (108 km) in a rocket-powered machine.

The Russian pilot remained inside the stratosphere, but the American climbed to beyond the mesosphere, far beyond the height to which a jet-powered aircraft can climb. A jet engine needs oxygen to burn its fuel, and at the stratopause, around 160,000 feet (48.3 km), with an air pressure of about 1 mb, there is not enough to sustain combustion. A rocket engine carries its own oxygen — it is not 'air-breathing' — and can operate outside the atmosphere. As you approach the stratopause, the absorption of incoming radiation raises the temperature to about 32 °F (0 °C) or even a little higher but, once in the mesosphere, you will find them falling again. By the time you reach 260,000 feet (79.3 km), the thermometer will show about -130 °F (-90 °C).

This is the height at which you may see noctilucent clouds above you. They are real clouds, although very tenuous, and made from water vapour. The water is a product of the breakdown of atmospheric methane, and the nuclei on to which it condenses enter the atmosphere from space — they are meteoric particles.

The stars should be clearly visible now.

(Opposite) Balloons are used extensively to measure conditions in the upper atmosphere. They are tracked from the ground, providing information about winds, and carry a variety of instruments.

The Lockheed TR-1 is designed to fly at very high altitudes, well inside the stratosphere, and is used for atmospheric research and aerial photography.

As you enter the mesopause, at about 300,000 feet (91.5 km), the air pressure is only 0.01 mb. You now need to protect yourself against the radiation bathing you from space and you should be wearing a spacesuit of the kind astronauts wear for working outside their spacecraft. You will not be 'weightless', like an astronaut. Although gravity will affect you less than it does at ground level, the reduction will be too small to make any appreciable difference to you. Astronauts do not lose their weight. They orbit the planet in such a way that their forward speed balances the rate at which they are falling downwards. They are in a free fall, but their forward motion prevents them from descending.

You will leave the mesopause — a region rather than a distinct boundary — at around 330,000 feet (100 km) and enter the thermosphere. The composition of the air is now changing. Bombarded incessantly by radiation, molecules of nitrogen and oxygen are separating, and the higher you go, the more these gases exist in their atomic, rather than molecular, forms. As you move still higher, more and more of the atoms are stripped of their electrons. The gases become ionized. The temperature is also rising. At 220 miles (354 km) it is about 1815 °F (990 °C), but the air is much too thin for you to feel it. Eventually, somewhere between 300 and 450 miles (483-725 km), you drift imperceptibly into the exosphere where the atmosphere consists of oxygen, hydrogen — produced from the breakdown of methane — and helium, produced by the exposure of nitrogen to intense radiation. About 1 per cent of all the atoms are ionized, and the lighter of them are escaping to join the solar atmosphere. If you climb above about 1200 miles (1900 km), your instruments will detect no gases as such, only free electrons and protons. You have entered the magnetosphere, and it is no longer possible to distinguish between the atmospheres of Earth and the Sun. At this great height they are united.

If there were some way in which the whole atmosphere could be compressed, so that its pressure and temperature were at a constant sea-level value throughout, all of the ozone in the air would comprise a layer no more than 0.12 in (3 mm) thick. Ozone is a rare gas. In the troposphere it occurs

The space shuttle (top) orbits the Earth at a height where the air is so thin as to be virtually — but not quite literally — non-existent. In this hostile environment, astronauts (bottom) must wear elaborate clothing to provide their bodies with insulation and air.

(Opposite) Radio waves used to be transmitted over the horizon by reflecting them off the ionized upper atmosphere. Today, they are more commonly relayed by an orbiting satellite which transmits signals that are received by sensors like this one (opposite) at Jodrell Bank, England.

naturally, but also as a pollutant and in the stratosphere, where it is said to protect us from over-exposure to solar radiation (see page 159).

Ozone is a form of oxygen. Oxygen atoms (O) usually combine in twos to form molecules (O_2). The ozone molecule consists of three oxygen atoms (O_3). This is not the preferred arrangement and, therefore, ozone is very unstable. It will react readily with any substance capable of oxidation and, for this reason, it will damage living tissue. This makes it poisonous to air-breathing organisms at quite low concentrations. Its three molecules are also separated easily by quite a small amount of energy.

If you can collect enough of it to make it visible, you will see that ozone is a pale blue gas, like oxygen. Unlike oxygen, however, it has a very distinctive smell. You have probably smelled it — and more than a trace of it will set you coughing and choking — if you have stood close to electrical apparatus that is sparking. Indeed, this is how the existence of ozone was first discovered, by M. van Marum in 1785, although it was not until 1840 that the German-Swiss chemist Christian Friedrich Schönbein (1799-1868) found that the smell was caused by a gas. It was Schönbein who named it, using the Greek word *ozon*, from *ozo*, meaning 'smell'. Lightning, which is a giant electrical spark, produces small amounts of tropospheric ozone, and ozone is also a product of the reactions, using energy supplied by sunlight, that cause photochemical smog.

Radiation from the Sun

The Earth is bathed in radiation from the Sun. This enters at wavelengths spanning a broad spectrum (see page 138), but it is most intense in the wavebands we perceive as visible light. Just beyond the violet end of the visible spectrum, at wavelengths between 0.1 and 0.5 micrometre, the radiation is called 'ultraviolet' (UV). About 9 per cent of the total solar radiation reaching the top of the atmosphere lies in the UV band.

As the incoming radiation encounters molecules of air, the collisions cause photons — the 'particles' that constitute electromagnetic radiation — to transfer some of their energy to the molecules they strike. This causes the radiation to be absorbed by the molecules. Whether or not it will be absorbed depends on a relationship

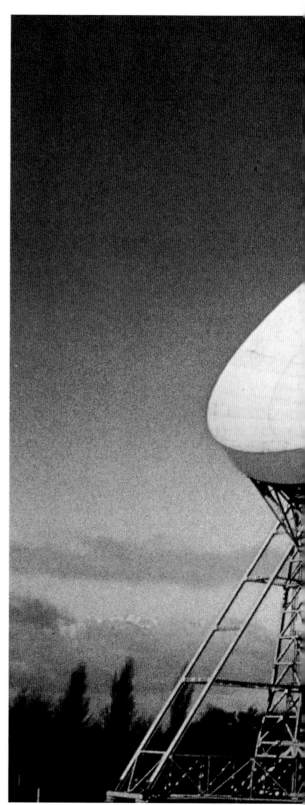

between the wavelength of the radiation and the size and shape of the molecule and, therefore, molecules of different substances absorb at their own particular wavelengths.

When UV radiation strikes oxygen molecules, the energy imparted to them breaks the bond linking them, and they

break into single oxygen atoms ($O_2 \rightarrow O + O$). Should one of these atoms then collide with an intact oxygen molecule it will bond to it, forming a molecule of ozone ($O + O_2 \rightarrow O_3$). Should an ozone molecule encounter a single oxygen atom, however, the ozone breaks apart and two ordinary oxygen molecules form ($O_3 + O \rightarrow 2O_2$). Ozone molecules are also split by UV radiation, at a slightly longer wavelength than those which break oxygen molecules. For as long as the atmosphere is exposed to UV radiation — which means throughout all the daylight hours — oxygen and ozone are engaged in this continuous process of formation, breakage and reformation, all the time absorbing the radiation that drives the reactions. The absorption of radiation at these wavelengths prevents it from penetrating further, thus confining the process to the upper air.

Radiation entering from outside the atmosphere will penetrate until it encounters a molecule that absorbs it. At altitudes higher than about 50 miles (80 km), the radiation level is intense, but the air is so rarefied that encounters with molecules are uncommon. Some UV is absorbed, raising the temperature as the energy imparted to them allows particles to move faster, but most penetrates deeper, to be absorbed mainly between about 20 and 40 miles (32-64 km).

In the upper part of this region of the stratosphere, above about 25 miles (40 km), the production and destruction of ozone are approximately in balance, but there is a certain amount of vertical movement, causing some mixing between air at this level and the air below. This mixing transports a little of the ozone to a lower level. There it is more protected from destruction by UV because the UV that might break it down is being absorbed at a higher level. This partial protection causes the ozone to accumulate between 12 and 22 miles (20-35 km), and its concentration is greatest at about 14 miles (22 km). This is the 'ozone layer'. It absorbs some short-wave UV radiation, but most is absorbed at higher levels. The accumulation of ozone in a distinct layer indicates that UV radiation is being absorbed, but not especially in the layer itself.

The shorter its wavelength, the more energy electromagnetic radiation possesses and, beyond the UV wavelengths, the Sun radiates in the most energetic part of the spectrum — it emits X-rays and gamma rays. These also split molecules into single atoms but they — and also UV to a lesser extent — go further. They can impart sufficient energy to the electrons that surround atomic nuclei for the electrons to break free.

An atom consists of a central nucleus, containing protons and neutrons, surrounded by electrons. The simplest, and lightest, atomic nucleus is that of hydrogen, which consists of just one proton. Protons carry a positive electrical charge, neutrons are electrically neutral and, therefore, an atomic nucleus carries a net positive electrical charge. (The number of neutrons affects the mass of the atom but, because they carry no charge, they do not affect its chemical characteristics.) An electron carries a negative electrical charge which is precisely equal to the positive charge on a proton. The number of protons in the nucleus determines the number of electrons that surround it, and a whole atom is electrically neutral. If the atom should lose electrons, however, it will be left with a net positive charge and the liberated electrons will have a negative charge. Such charged particles are said to be 'ionized' and the particles themselves are 'ions'.

Above about 40 miles (64 km) the air is increasingly ionized, with the greatest concentration of ions between 40 and 200 miles (64-322 km). Above this height, the air is completely ionized but the concentration of ions is lower because the atmosphere is increasingly tenuous. This region, above about 50 miles (80 km), is called the 'ionosphere'. It is where aurorae occur (see page 144) — although these can form at heights up to more than 600 miles (965 km).

Discovery of the ionosphere

Changes he observed in the Earth's magnetic field led the eminent physicist, Balfour Stewart, to propose the existence of such a layer, in about 1882, but the real discovery followed the first successful transmission of a broadcast signal across the Atlantic, by the Italian physicist, Guglielmo Marconi (1874-1937), in 1901. Until then, it had been supposed that radio waves, travelling in straight lines, could not be received beyond the horizon because they would disappear into space at a tangent from the Earth's surface. Their reception at a much greater distance suggested they were being reflected, and this thought led two scientists, working independently of one another, to a similar conclusion. Arthur Edwin Kennelly, working at Harvard University, and Oliver Heaviside (1850-1925), who had been forced by deafness to retire from his job with a telegraph company and who worked alone in Devon, England, suggested that the radio waves were being reflected by an electrically conducting layer in the upper atmosphere. For a long time, this layer was called the 'Kennelly-Heaviside layer'.

The new name, ionosphere, was adopted

(Opposite) A hot-air balloon flies because the heated air is less dense than the surrounding air and tends to rise over it.

around 1930. By then the properties of the layer had been explored in some detail. It had been found that high frequency radio waves, at more than 15-20 megahertz (million cycles per second) pass through the layer, and those at 1-2 megahertz are either reflected or absorbed by it. The shorter the wavelength of the transmission, the greater the density of electrons that is required to reflect it. The altitude of the layer was measured, in 1925, by reflecting radio waves from it and measuring the time that elapsed between the transmission and the reception of the reflected signal. More recently, the upper side of the ionosphere has been examined by satellites.

Until communications satellites came into general use for receiving and relaying long-distance transmissions, it was the ionosphere that allowed people to receive broadcasts over thousands of miles. Today, it is still the ionosphere that determines which radio wavebands can be used, because only those that can penetrate will be received by and from orbiting satellites and, even then, the ionosphere is able to interfere with radio transmissions. During periods of intense solar activity, when the solar wind 'blows' strongly, ionization increases and more radio wavelengths are absorbed. This causes the signal to fade and, in extreme cases, to be lost altogether, in a 'radio black-out'.

In May, 1958, readings from a Geiger counter he had installed on the satellite *Explorer 1* led James A. Van Allen to report his discovery of two bands of intense radiation, at 620 to 3100 miles (1000-5000 km) and 9300 to 15,500 miles (15,000-25,000 km), forming a doughnut shape above the Equator (they are virtually absent above the poles), where charged particles move, in approximately corkscrew paths, along the Earth's magnetic field. The bands were named the Van Allen radiation belts. The lower band contains protons and electrons, the higher band mainly electrons, and all of them are highly energetic. Scientists believe that, although the testing of nuclear weapons contributed to them and some arrive from outside the solar system as part of the cosmic radiation, most of the particles are part of the solar wind — and, therefore, of the solar atmosphere — that have been trapped by the magnetic field. The radiation — through which astronauts must travel, although they do not remain

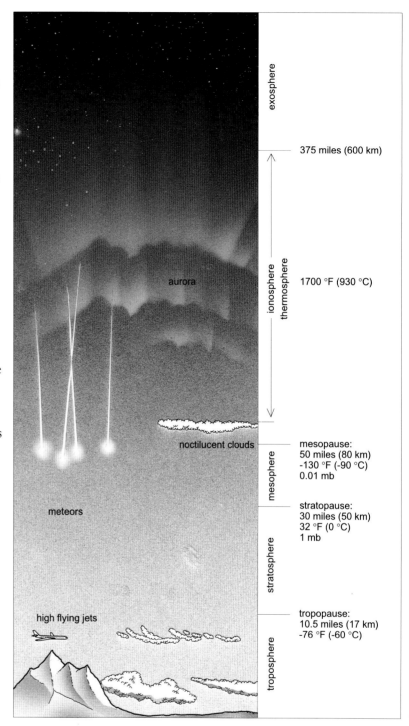

exposed for very long — is caused by collisions between incoming radiation and atmospheric atoms. This produces neutrons, some of which decay into highly energetic protons and electrons — the protons have enough energy to penetrate several inches of lead. In the outer belt, increases in the numbers of electrons, and sometimes of protons, are associated with magnetic storms. The inner Van Allen radiation belt is where the atmospheres of the Earth and Sun meet and merge.

A visual 'model' of the Earth's atmosphere.

(Opposite) The central United States has a typically continental climate. It is relatively dry, with warm summers and cold winters.

HOW AIR MOVES

If all the Sun's radiation that reaches the Earth was absorbed in the atmosphere, life would need to be organized very differently — indeed, if life were possible at all. There could be no photosynthesis, for example, and eyes would be of little use in a world of perpetual darkness. In fact, the atmosphere absorbs rather less than 20 per cent of the radiation falling upon it. Ozone absorbs much of the UV and water vapour absorbs in several bands between about 0.9 and 2.1 micrometres. Clouds, haze consisting of atmospheric particles, and light-coloured regions of the surface (see page 148) reflect an additional 10 per cent so that, of the radiation arriving at the top of the atmosphere, about 70 per cent penetrates all the way to the surface.

This radiation is absorbed by the surface and, when a body absorbs energy, it warms — the energy is transferred to its molecules which vibrate more rapidly — and then it, too, begins to radiate. This is called 'black-body' radiation, and a perfect black body absorbs all the radiation it receives and re-radiates all of that energy at the maximum rate possible — which depends on its temperature. The Earth is by no means a perfect black body but, nevertheless, it emits radiation at wavelengths beyond the red end of the spectrum (see page 141). Although the atmosphere is largely transparent to short-wave radiation, it is partially opaque to radiation at longer wavelengths and, therefore, some of the Earth's own, black-body, radiation is absorbed. (This is the 'greenhouse effect', described in more detail on pages 161-63.)

The air is warmed, therefore, by its own absorption of long-wave radiation and, where water vapour condenses, by the latent heat of condensation, but its main source of warmth is contact with the surface. The greenhouse absorption of black-body radiation also causes some warming of the surface, because the molecules that absorb long-wave radiation also behave as black bodies, radiating their own energy in all directions — some of it downwards.

Air is a poor conductor of heat but it is very mobile. This means that although its molecules are very inefficient at passing energy from one to another by direct contact, they are much more efficient at transferring heat by convection. When air is warmed by contact with the surface, it rises. If you picture it as a 'parcel' of air, as it rises it will cool adiabatically — fairly rapidly while it remains unsaturated and more slowly as its water vapour starts to condense and releases latent heat to offset the cooling. Adiabatic cooling and warming involve no mixing between the 'parcel' of air and the air surrounding it. This is the way rising air cools once it is well clear of the surface — although there is usually some mixing with the surrounding air — but close to the surface it is more usual for air to be mixed fairly thoroughly by local turbulence as it moves laterally across the uneven ground surface.

The Earth is warmed by the Sun more strongly at the Equator than at the poles. This suggests the possibility of a quite simple convection system. In low latitudes in each hemisphere, air is made very warm. It rises and its place near the surface is taken by cooler, denser air drawn in from a higher latitude. The rising air cools adiabatically and, somewhere near the pole, it is drawn downwards again to replace air that has moved equatorwards. Throughout the whole of the troposphere, the air is engaged in a kind of rolling motion.

If this is all that happened, we should expect water vapour, carried upwards in the warm, low-latitude air, to condense and fall as rain not very far from the Equator. In higher latitudes the air would be very dry — 'squeezed' dry as the air approached the tropopause — and so the climate of the tropics would be very humid and that of high latitudes very dry. As we know from our own experience of the weather from one day to the next, it is not so simple!

Heating air from below
The oversimplification begins with our implied assumption that the warming of the surface occurs evenly. It does not. Nearly three-quarters of the surface of the Earth is covered by water. Like air, water is a poor conductor of heat. It also has a high 'specific heat'. This is the amount of heat energy that a given volume of a substance must absorb to increase its temperature by a specified

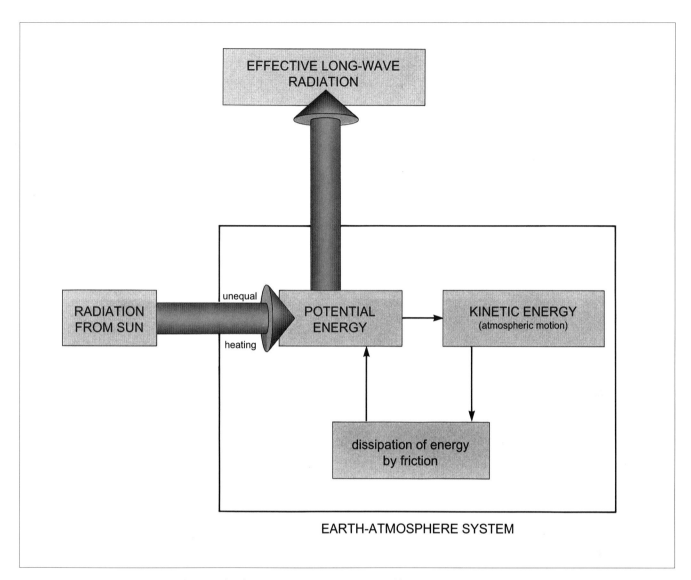

EFFECTIVE LONG-WAVE RADIATION

RADIATION FROM SUN

unequal

heating

POTENTIAL ENERGY

KINETIC ENERGY (atmospheric motion)

dissipation of energy by friction

EARTH-ATMOSPHERE SYSTEM

Energy changes in the Earth-atmosphere system

amount. Because its specific heat is high, it takes a considerable amount of solar radiation to alter its temperature. Unlike air, water is warmed from above, so a surface layer of warm water overlies the main water mass, which is much cooler. The two mix, transporting heat downwards from the surface, and the consequence is that the oceans warm and cool slowly.

Rocks, on the other hand, have a specific heat about one-fifth that of water. They, and soil particles, which are made from rock fragments, warm and cool quickly and heat can be transferred by conduction among particles that are in contact with one another. There is a very marked difference, therefore, between the effect of surface warming over land and over water.

There is also a difference between one soil and another. If the soil consists of large rock particles, loosely packed with air spaces between them, the air will act as an insula-

tor. The surface will warm rapidly during the day, but its heat will not be conducted to any great depth because of the air. Sand is composed of large mineral particles of this kind. Where the particles are smaller, as in a clay soil, for example, particles will be more closely packed, with smaller air spaces, and heat will be conducted more effectively, so the soil warms quickly, but also absorbs more heat than a sandy soil. In a moist soil, spaces between solid particles will tend to be filled, and the soil will be a better conductor. If the soil is very wet, however, the high specific heat of the water will become the dominant factor, and the soil will heat and cool more like water than like dry land.

Such variations, some very local and others vast, influence the way the surface responds to the warmth it receives. Even then, this qualification of what was clearly an oversimplified description contains yet another false assumption — that the surface

Satellite photographs reveal clearly how moving air is deflected by the Coriolis 'force'.

is evenly exposed to incoming radiation. On a local scale it is sometimes obvious that this is not so. In a steep-sided valley that is aligned approximately east-to-west, for example, one side is inclined towards the Sun (in the northern hemisphere, the south-facing side) and the other away from it, in perpetual shade. The bottom of the valley may also be shaded for most of the time. Where sunshine does fall evenly, some surfaces reflect it, so it cannot be absorbed. The reflectiveness of a surface is known technically as its 'albedo' (described more fully on page 148) and, in the case of water, its value can be very high or very low depending on the angle of the Sun. When the Sun is high in the sky, the sea looks very dark but, when it is low, the reflection from the surface of fairly calm water is almost total.

The most dramatic variation in reflectivity, however, is caused by clouds. Their presence or absence has a profound effect on the temperature at the surface. In extreme cases, big cumulonimbus storm clouds block almost all the solar radiation, turning day into night.

Our idea of a 'rolling motion' — in which the Earth's surface is warmed in low latitudes, warms the air above it, and the warm air rises, cools, and sinks again in high latitudes, — is very crude. The amount of sunshine reaching the surface varies according to the amount of cloud cover and the inclination of the surface — the local topography — and the extent to which it warms the air depends on the nature of the surface itself. The air is being warmed more strongly in low than in high latitudes, but the effect is very uneven. If the air is 'rolling', then it is doing so not as a single, homogeneous mass, but as many smaller masses, all moving at different rates.

The picture is more complicated, but still it does not describe what happens in the real world, because it overlooks the most obvious characteristic of any planet — its rotation about its own axis. This affects the movement of air in two distinct ways, one through the conservation of angular momentum, the other because of the Coriolis 'force'.

A body that is moving at a constant speed has 'momentum' — it will continue to move in the same direction at the same speed unless a force acts to alter the situation. Momentum is proportional to the mass of the body and its speed of motion. If a body is spinning on its own axis, each part of it has momentum, in this case 'angular momentum'. This is proportional to its mass, the square of its distance from the axis and its 'angular velocity'. This is its rate of rotation, measured as the number of degrees through which it passes in a specified length of time — in the case of the Earth, one complete revolution of 360 degrees in 24 hours and, therefore, 0.25 degrees, or 15 minutes of arc, per minute.

The total amount of angular momentum for all the parts making up the body must remain the same for as long as the mass of the body and its rate of spin remain constant. In other words, angular momentum must be conserved. This means that if one component of the equation for calculating angular moment changes, there must be a compensating change in some other component. There are only three components — mass, distance from the rotational axis and angular velocity. Presuming that the mass of the body is fixed, the only variables are the distance from the axis and the angular velocity. If one of these decreases, therefore, the other must increase.

The influence of the Earth's rotation

The Earth's atmosphere moves with the surface of the Earth, rotating as the planet rotates. If a mass of air moves from a low latitude into a higher latitude, however, its distance from the Earth's rotational axis has decreased. The axis of rotation forms a straight line through the centre of the planet, so the distance between it and a body on the surface is at a maximum at the Equator and zero at the (rotational) poles. One factor in the angular momentum equation decreases and the only factor that can, and does, increase is the angular velocity. As a mass of air moves away from the Equator it is accelerated in the direction of the Earth's rotation eastwards (in both hemispheres).

The Coriolis 'force' was discovered, in 1835, by a French engineer, Gaspard Gustave de Coriolis (1792-1843). It explains why an object moving away from the Equator will appear to be deflected to the right of its direction of movement in the northern hemisphere and to the left in the southern. The effect is proportional to the latitude and the horizontal speed of the moving object but it is not really a force at all. It is not the body that is deflected, but the Earth beneath it which has moved, because of its rotation, while the object was passing over it. You can demonstrate the effect with a disc, pivoted through its centre. Rotate the disc slowly and, while it is rotating, try drawing a straight line from the edge towards the centre. When the disc stops you will see the line is curved — but only because of the rotation of the disc.

To either side of the Equator, the prevailing winds blow from the east — from the north-east in the northern hemisphere and from the south-east in the southern. Two hundred years ago, when international commerce relied on ships powered by wind, these reliable winds were of great importance. Mariners called them the 'trade winds' and they have kept their name. They were also of great scientific interest. What made them blow so dependably?

The Hadley cells

In 1735, a London lawyer and part-time scientist called George Hadley believed he had found the answer, and presented it as a paper to the Royal Society of London. Hadley proposed that warm, equatorial air rises, moves away from the Equator as it does so and cools at a great height. Its place near the surface is taken by cooler air, drawn in from a higher latitude, and so the air

The Coriolis 'force'. (Top) If you try to draw a straight line from the centre to the circumference of a rotating disc, the resulting line will be curved because the disc moved beneath the pencil while it was being drawn. (Centre) Also on a rotating disc, suppose a person at X, a short distance from the centre of the disc, sees an object at T, on the circumference. The line of sight is straight. (Bottom) If the person at X tries to throw a ball in a straight line to T, however, the ball will follow an apparently curved path to land at Z, because during its flight position X moved to X+ and position T moved to T+. No 'force' is involved; the phenomenon is due only to the rotation of the disc.

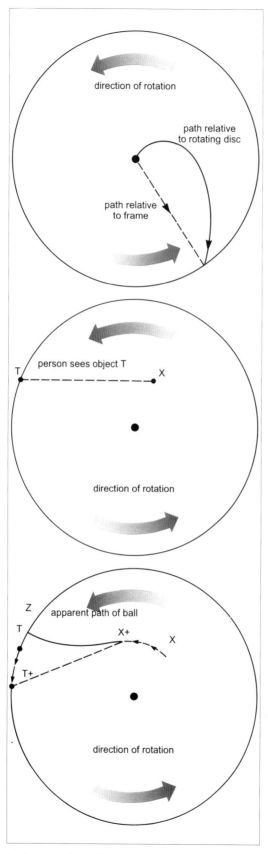

occupy the whole of each hemisphere — the cooling air descends in the subtropics. As the air returns to the Equator, the rotation of the Earth deflects it, so the winds blow from the north-east and south-east, rather than directly from the north and south. Hadley thought the deflection was due to what came to be called the Coriolis 'force' but, in fact, the conservation of angular momentum is probably more influential — it accelerates objects in an easterly direction when they move away from the Equator, but in a westerly direction (an 'easterly' wind blows from east to west) when they move towards the Equator.

The Earth spins on its axis at a constant rate. This means that, over the world as a whole, the energy that is expended in driving easterly winds must be balanced by a force acting in a westerly direction. If this were not so, the Earth would be slowing down and we would need to replace our clocks and watches at regular intervals, because of the steady increase in the length of the days. The balancing westerly winds occur in the middle latitudes.

The movement Hadley described forms what are now known as 'Hadley cells' and they are driven by convection. They are rather more complex than they sound. In the first place, their centre is not the geographical Equator, but the thermal Equator — the line around the Earth joining points of highest average temperature. Its position varies from time to time but, on average, it lies at about 5 °N. The cells are not continuous right around the Earth, partly because the surface is not heated evenly and, despite their reputation, the trade winds do not blow everywhere, all the time. Pairs of Hadley cells, on either side of the Equator, occur only in spring and autumn and the cell that forms in winter — first on one side of the Equator, then on the other — is the strongest of them all, some of its air crossing the Equator at a low level, into the summer hemisphere.

In summer, the Hadley cells help to produce the monsoon rains. The mechanism is complicated, and linked to changes in the pattern of high-altitude winds and the jet streams (see page 90), but essentially it is convective, of the Hadley-cell type.

Land warms more quickly than the oceans and in summer, therefore, the air over the continents is warmer and less dense than the air over the sea. This reduces the

'rolls', much as the simple description outlined earlier, based mainly on common sense, would suggest. The difference is that the motion described by Hadley does not

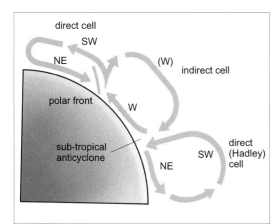

atmospheric pressure over the continents, especially over north-western India. Condensation in the rising air leads to spectacular cloud formations and dramatic storms, but the land is arid after its dry winter and, therefore, the total rainfall is slight. The weather becomes very hot — between March and May, the average temperature in Delhi increases from 74 °F (23 °C) to 92 °F (33 °C) — and there are dust storms. The surface atmospheric pressure continues to fall until there is a very marked difference between the low pressure over north-western India and high pressure over the Indian Ocean to about 10 °S — in the southern hemisphere.

Then, by around the end of May, the gradient of pressure between the continent and the ocean becomes so strong that air is drawn from the high-pressure region, across the Equator, to fill the continental low-pressure region. This air is also warm and it is very humid, carrying water that has evaporated into it from the Indian Ocean and as it crosses the Arabian Sea.

The air arrives, in late May and early June, as south-westerly winds. As they cross into western India they encounter high ground which forces the air to rise — this is called 'orographic lifting'. As the air rises it cools, its water vapour condenses, and intensely heavy rain falls along the western side of the country and there are fierce storms, caused by convection, further inland. Local areas of low pressure — depressions — associated with this weather system then move westwards across India, bringing rain to more distant regions. The weather in early summer is so predictable, and so dramatic, that Arabic speakers have called it *mawsim*, which means 'fixed season', the word we have altered to 'monsoon'. Strictly speaking, it refers to the

whole season, not simply the rains. We tend to associate the monsoon with India, but it also occurs in West Africa, southern China and northern Australia.

The air in the Hadley cells descends at about latitude 30°, to form areas of high pressure in the subtropics. Of the world's total evaporation of moisture, 60 per cent occurs between 25 °N and 25 °S and 60 per cent of that moisture is 'squeezed' out of it, condensing and falling in the humid tropics. Not surprisingly, the subsiding air is associated with the belts of deserts that circle the world in the subtropics of both hemispheres.

In middle latitudes, beyond the easterly trade winds, the prevailing winds blow from the west — mainly from the south-west in the northern hemisphere and the north-west in the southern. In general, they blow away from the subtropical areas of high pressure and towards the areas of low pressure which surround both poles. Areas of high pressure lie above both poles, surrounded by belts of mainly easterly winds.

The Coriolis 'force' exerts no effect at the Equator, but its influence increases with increasing latitude. For this reason, it is only in low latitudes that the circulation of the atmosphere is governed mainly by convection. The overall circulation leads, nevertheless, to the transfer of energy away from the Equator and towards the poles, and two more systems of cells, much weaker than the Hadley cells, occur in higher latitudes, one direct, the other indirect, in that it is driven by the cells to its north and south. This is known as the 'three cell model'.

The subsidence of very cold air, which produces the area of high pressure at each pole, causes an easterly outward flow from polar into subpolar regions. There, at about latitude 60°, the low-level cold air mixes with middle-latitude air and forces it to rise. The subsiding air of the Hadley cells marks a region of general air subsidence, and so the combination of air moving away from the polar regions in very high latitudes, and moving towards the Equator in the tropical cells drives a middle-latitude system of cells. This very general outline greatly simplifies a pattern that is subject to countless local variations — and about which much remains to be discovered. It is no more than a summary of the overall effect, and you should not regard it as an accurate description of what is happening in any particular place or at any particular time.

Global atmospheric circulation

A line, drawn on a map, which links points where the atmospheric pressure is the same is called an 'isobar', and a set of isobars will outline regions of relatively high and relatively low pressure. The resulting map looks much like a conventional contour map, where the contour lines link places which are at the same height above sea level. Indeed, the similarity is so striking that isobars are often called 'contours' and pressure maps 'contour maps'. If you regard a pressure-distribution map as though it were a real contour map, regions of high pressure become hills or ridges and regions of low pressure are depressions or valleys. 'Ridge' and 'depression' are words you will often hear in weather forecasts, but valleys are called 'troughs'. This is what the words mean. If you extend the analogy, the movement of air becomes equivalent to the movement of water over the land — it flows down slopes or, abandoning the analogy, from high pressure to low pressure. For rather similar reasons, at high altitudes air also flows from regions of high to low temperature. You can see why this is so when you remember that cold air is denser than warm air. Being more compact it occupies a smaller volume and so, returning to the analogy with contours, its upper surface lies at the lower level — the air flows 'downhill'. Moving air forms a wind, and a wind flowing from warm to cold air is called a 'thermal wind'.

At this point the analogy breaks down, because air does not flow between restraining banks and it is subject to the Coriolis 'force'. The air tends to flow in a straight line from the centre of an area of high pressure to the centre of an area of low pressure — at right angles to the isobars. The Coriolis 'force' acts at right angles to the direction of air movement — parallel to the isobars. Near the surface, the result is that the air flows across the isobars at an angle. High above the surface, however, the Coriolis 'force' wins, and the air flow, in what is called the 'geostrophic wind', is parallel to the isobars. Seen from above, in the northern hemisphere the air flows anticlockwise around areas of low pressure and clockwise around areas of high pressure — stand with your back to the wind and the area of low pressure is to your left. In the southern hemisphere these directions are reversed. This was first stated by Professor Buys Ballot, of Utrecht, in 1857, and it is known as 'Buys Ballot's law'.

It is the differences in pressure — the pressure gradient — which drive the geostrophic winds and the differences in temperature — the temperature gradient — which drive the thermal winds. The thermal wind is the principal source of the westerlies which prevail in middle latitudes in both hemispheres. Temperature is highest at the Equator, lowest at the poles and, allowing for local disturbances, it changes at a fairly steady rate as you move between the two. There is a tendency, therefore, for a thermal wind to blow directly from the tropics to polar regions. As it does so, it is deflected by the Coriolis 'force' to the right in the northern hemisphere and to the left in the southern hemisphere. In both hemispheres the deflection adds an easterly component to the direction of flow, producing south-westerly (blowing from south-west to north-east) winds in the northern hemisphere and north-westerlies in the southern hemisphere. At high altitude, however, the thermal winds become virtually geostrophic, flowing parallel to the temperature gradient rather than at right angles to it.

The general distribution of pressure amplifies the westerlies. The high pressure of the subtropics adjoins the low pressure of the subpolar regions. In the northern hemisphere, air moving in a generally anticlockwise direction is flowing from west to east on the southern side of the subpolar low. At the same time, air moving in a clockwise direction around the subtropical high is also moving from west to east at its northern boundary. In the southern hemisphere, the boundary is between the southern side of the subtropical high, where the anticlockwise movement produces a west-to-east flow, and northern side of the subpolar low, where the clockwise movement also produces a west-to-east flow.

At this point you may begin to suspect that something is seriously wrong with the general description because, although westerly winds are familiar enough to people living in middle latitudes, calling them 'prevailing' seems to stretch the point. In Britain, for example, the wind can and does blow from every direction. Surely, you may think, it is unrealistic to assert that the wind blows from the west more often than from any other direction. If you measure the direction of the wind every day, year after

(Opposite) The tropopause forms a barrier that rising tropospheric air cannot cross. The 'anvil' at the top of high clouds marks the upward limit of vertical motion. Such convective clouds are common close to the Equator, where they demonstrate the movement of air vertically and away from the Equator within the Hadley cells.

year, however, you will find that it blows from a westerly quarter more often than it blows from any other quarter — though not necessarily more often than from all other quarters combined.

There would be no doubt at all about the prevalence of westerly winds if the distribution of pressure was more orderly. As it is, however, the regions of generally high and generally low pressure tend to form local cells rather than belts. The overall pattern reflects the distribution of solar energy according to latitude, but air is warmed and cooled by its contact with the ground and this also influences the distribution of pressure. In summer, the strong heating of the surface in the interior of continents produces regions of low pressure, while higher pressure prevails over the oceans. In winter, the situation is reversed as the oceans retain their heat while the continents cool rapidly.

To complicate the picture further, the middle latitudes are where the fairly stable tropical-subtropical and polar-subpolar regimes meet. The result is a patchy, irregular mixture of regions of high and low pressure. The whole mixture moves, as a set of weather systems encircling the globe. The movement is always in an easterly direction (from west to east) in both hemispheres, because it is influenced by the conservation of angular momentum acting on air that is moving from lower to higher latitudes, down the thermal gradient.

The air does not all move at the same speed. In summer, for example, a large area of high pressure often becomes established somewhere near the Azores and may remain stationary for several weeks. Air moving across the North Atlantic is deflected around it, to the north or south. When this happens, north-western Europe, including Britain, falls under the climatic influence of the Azores high, which brings warm, dry, sunny weather — and when the high proves especially stubborn, drought. Should the summer high fail to become established, there is nothing to deflect the movement of the weather systems and the summer is more likely to be cool, cloudy and wet.

At a more local level still, the movement of air over land, and the weather it brings, are strongly influenced by the 'roughness' of the ground. Friction reduces the speed of the wind and large objects may reduce it further. It is generally windier in rural areas than in large cities. Orographic lifting — the passage of air over hills — causes condensation and precipitation on one side of the high ground, but the other side, in the rain 'shadow', remains dry.

Unusual weather patterns

In the course of about two weeks in January, 1987, some 2000 people died in Britain because their homes were inadequately insulated and they were too poor to heat them. They were unable to withstand the unusually cold weather which, for that short time, gripped western Europe. People described the conditions as 'Siberian'. They were right, but the cause of the phenomenon was curious.

The air over western Europe, and the weather associated with it, arrive from the west, having crossed the Atlantic. The air is moist and its temperature is moderated by its prolonged contact with the ocean surface. In winter, when the sea-surface temperature is higher than the surface temperature of dry land, the air feels mild. As it continues to move eastwards across the continental land mass, it loses much of its moisture, it is in contact with the cold ground, and its characteristics change. It becomes drier and colder. By the time it reaches central Asia — Siberia — it is intensely cold, very dry and, therefore, extremely dense. It settles as a large area of high pressure, with its centre somewhere near Novosibirsk, and it brings truly Siberian conditions to the region it covers. We know that the North and South Poles are cold, but the coldest places on Earth are not located at the poles themselves. In the northern hemisphere, the place with the coldest winters — the 'cold pole' — is Verkhoyansk, in north-eastern Siberia and only just inside the Arctic Circle, where the average January temperature is about -58 °F (-50 °C) and it can go down to -90 °F (-68 °C). (Cold though this is, Antarctica is colder. In some places there a temperature of -130 °F (-90 °C) has been recorded.)

Air will tend to spill out from such a large, dense mass. As the air tries to spill westwards, however, it is constrained by the warmer air that is moving eastwards, towards it. The milder, less dense air rides up, over the denser air — much as moving air is forced to rise up the side of a hill — but, in doing so, it also exerts a pressure against the cold air. The vigour with which the advanc-

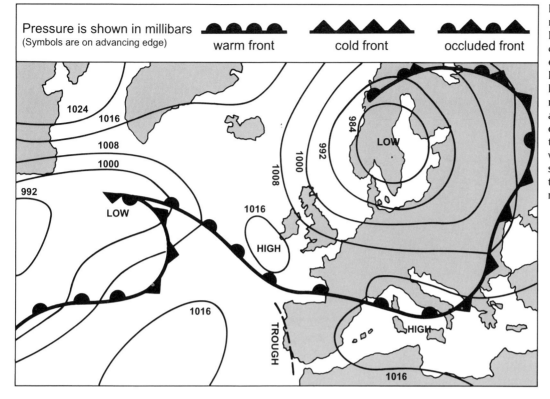

Pressure is shown in millibars (Symbols are on advancing edge)

warm front   cold front   occluded front

1024
1016
1008
1000
992
LOW
1016
HIGH
1016
TROUGH
984
LOW
1000
1008
992
1000
1008
1016
HIGH
1016

Blocking highs and moving lows in the North Atlantic. The high over the Azores, with an extension to the west of Britain, is stationary. The low over Scandinavia is moving eastwards around the high, and the entire weather system in the western Atlantic, with low pressure to the south and high pressure to the north, is moving northwards.

ing air rises, and the pressure it exerts, are proportional to the difference in temperature between the two masses. The colder the cold air, or the warmer the mild air, the more rigorously the cold air is held in place.

In the winter of 1987, the weather over Siberia was unusually mild — for Siberia, of course. The central Asian air mass was not quite so cold as it is in most years and, therefore, the difference in temperature between it and the air to its west was smaller than usual — the temperature gradient was shallower. The cold air was not held back so firmly, and some of it managed to spill westwards as far as the Atlantic coast. Europeans shivered but, paradoxically, the low temperature they experienced was caused by the relatively warm conditions in Siberia. Some climatologists suggested the warmth might have been a symptom of the 'greenhouse effect' (see page 161). No one can say whether this is the case, but it demonstrates that a global warming does not necessarily mean everyone will enjoy warm weather all the time.

As air moves across the surface of the Earth, its character is altered by the conditions it encounters to an extent that is determined by the length of time it remains within a particular environment. It may be warmed or chilled and it may acquire water vapour by evaporation or lose it by conden-

sation and precipitation. When the temperature, humidity and lapse rate of the air are more or less uniform over hundreds of square miles, the air is said to constitute an 'air mass'. The weather we experience from day to day is determined to a very large extent by the air mass which produces it.

Air masses and fronts

An air mass is necessarily large — this is not merely a conceptual convenience or convention. A small body of air cannot remain isolated for very long from the air surrounding it. It will mix with the larger body of air and the mixing — of all small 'parcels' with their surroundings — will homogenize the air fairly efficiently. In other words, if you begin with small, local bodies of air, all of them different, before long the differences will be eliminated by mixing and the result will be a large, fairly homogeneous body — an air mass.

Different types of air masses — warm, cool or cold, moist or dry — are given names which are related to the kinds of areas in which they acquired their characteristics. The descriptions are very broad and there are not many types but the labelling is useful because the names immediately suggest the conditions likely to be associated with them. Each name consists of two words and they are usually abbreviated

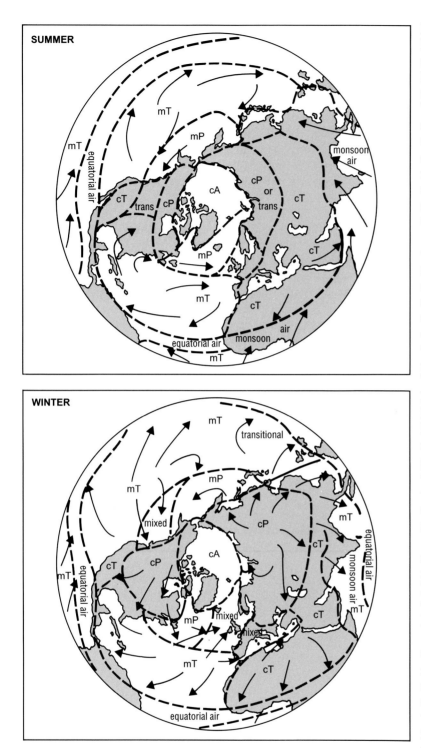

SUMMER

mT

mP

mT

equatorial air

cT

trans

cP

cP

cA

cP or trans

cT

monsoon air

cT

mP

mT

cT

equatorial air

monsoon air

mT

WINTER

mT

transitional

mT

mP

mixed

mT

cP

cT

mT

cT

cP

cA

cT

equatorial air

monsoon air

mT

mP

mixed

cT

mixed

mT

cT

equatorial air

Air masses in the northern hemisphere

mA = maritime arctic
mT = maritime tropical
mP = maritime polar
cT = continental tropical
cP = continental polar
cA = continental arctic
trans = transitional

to initial letters — two letters for the name of each type. By convention, the initial letter for the first word of the name is written in lower case and the second as a capital, so names have forms like 'cT', 'mP' or 'cP'.

The 'c' and 'm' stand for 'continental' and 'maritime'. They describe air in which the characteristics have been acquired over a large continental land mass or over the ocean. Continents and oceans occur in all

latitudes so, by itself, the designation tells us little. It is qualified by the second part of the name, which refers more directly to the latitude and, therefore, the temperature of the region from which the air mass came. Written with a capital letter when used alone, this may be arctic (A), polar (P) or tropical (T). These five letters can be combined to label all six of the air mass types — continental arctic (cA), continental polar (cP), maritime arctic (mA), maritime polar (mP), continental tropical (cT) and maritime tropical (mT).

In addition to these, equatorial air, within the convectional Hadley cell regime, is considered separately, as is monsoon air during the monsoon period. Where one type of air mass is changing into another — as continental air moves across an ocean, for example — there are transitional areas in which the air is of an intermediate type and the type name becomes an unreliable guide. There are also regions — north-western Europe is the largest — in which the boundaries of air masses meet and move back and forth over short periods. This makes it impossible to describe the region in general terms, because sometimes it lies within one air-mass type and sometimes within another. At different times, for example, Britain may be covered by mP, cA, mT, cP, or cT air.

The geographical area within which an air mass originates is called its 'source region'. In the northern hemisphere, the high-pressure systems over the Arctic and the northern parts of North America and Eurasia are the sources of cold continental (cA and cP) air, and the North Atlantic and North Pacific are the sources of cold maritime (mP) air. In the southern hemisphere, Antarctica is the source of cold continental (cA) air, and the seas surrounding that continent the source of cold maritime (mP) air. The sources of tropical air are at lower latitudes — but they are not confined to the tropics. Continental tropical (cT) air forms over the southern part of North America, and in a large region extending across Africa north of the tropics, the Arabian Peninsula and into Asia — where it is sandwiched between the cP source region to the north and the monsoon region over India. The low-latitude oceans are the sources of maritime tropical (mT) air. In the southern hemisphere, cT air originates over southern Africa, Australia and, but to only a small extent, over South America, and the southern oceans between the equatorial air to the north and the boundary of the mP air to the south are the source of mT air.

When water evaporates, the latent heat of evaporation cools its surroundings — this is the physical principle our bodies exploit when the evaporation of sweat from our skins cools us. As water vapour condenses, the latent heat of condensation warms the surrounding air. Taken together, the latent heats of evaporation and condensation act to moderate air temperatures and, therefore, temperatures are likely to be more extreme when the air is dry. Continental air, originating far inland, is dry. In summer, as the ground surface warms, it becomes hot and the boundary between the continental tropical and continental polar air masses shifts polewards. In winter, as the land surface cools, it becomes very cold and the boundary shifts towards the Equator. Continental air brings hot, dry weather in summer and cold, dry weather in winter.

Maritime air, originating over the ocean, is moist, the amount of water it carries depending on its temperature — mT air is moister than mP air. Because of its water vapour, the temperature of the air is moderate in comparison with that of continental air. In summer, maritime air is associated with warm temperatures and rain, in winter with mild temperatures and rain or snow.

Consider what happens, however, as an air mass of either type moves away from its source region. When relatively cool air moves across a relatively warm surface, the air will be warmed from below. This will cause air to rise by convection and, if there is a large difference between the temperatures of the air and surface, air may be lofted to a great height. Such moisture as it contains will condense to form clouds and, because these are associated with strong convection currents — making the air 'unstable' — the clouds will be of the cumulus, or heap, type. In the case of maritime air that crosses a warm land mass in summer, the clouds are likely to be large and numerous, and there may be showers and, perhaps, storms. These weather conditions will last for as long as the air retains sufficient water vapour but, as it moves inland, it continues to lose moisture until, eventually, it has been thoroughly dried and warmed. Then it will have become an air

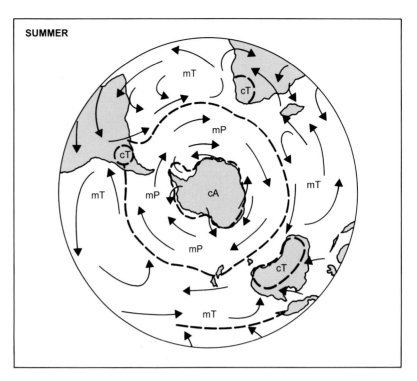

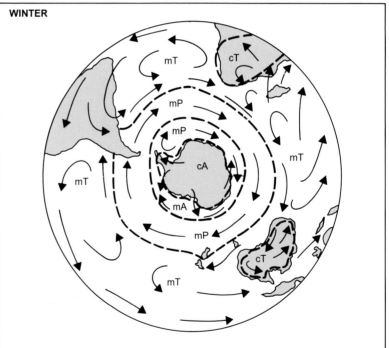

mass of a different type.

Air can also be cooled from below, when warm air crosses a cooler surface, and this produces a very different situation. The layer of air that is in contact with the ground becomes cooler and, therefore, denser. It tends to subside, while warmer, less dense air lies above it. This is a 'temperature inversion'. There is little or no vertical movement and the air is said to be 'stable'. If the air is of a maritime type, the

Air masses in the southern hemisphere

mA = maritime arctic
mT = maritime tropical
mP = maritime polar
cT = continental tropical
cP = continental polar
cA = continental arctic

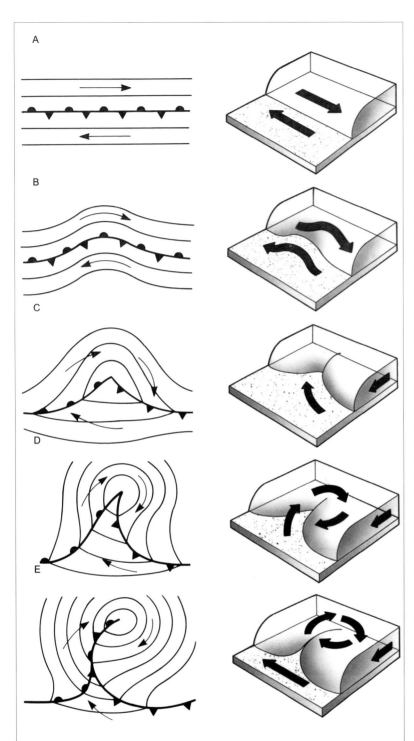

A

B

C

D

E

typical of its source region. When it moves away from its source region, however, the weather associated with it is modified strongly by the surface across which it travels. The weather is also modified where air masses of different types collide.

All air masses move, but they do not all move at the same speed, so collisions are common. The boundary between two air masses moving at different speeds is called a 'front'. Cold air is denser than warm air and so, for equal volumes of air, the pressure it exerts at the surface will be higher — it forms areas of high pressure, known technically as 'anticyclones'. Warm air, being less dense, forms 'cyclones' — areas of low pressure. In each case, the air is 'warm' or 'cold' only in relation to the adjacent air and not in any absolute sense. As it moves to equalize the pressure, (in the northern hemisphere) air travels clockwise ('anticyclonically') around centres of high pressure, and anticlockwise ('cyclonically') around centres of low pressure. The forces which govern the motion of air in a curved path on the spinning Earth combine in such a way that an air mass which is moving anticyclonically in the northern hemisphere and cyclonically in the southern hemisphere travels faster than one moving in the opposite sense. The consequence is that high-pressure areas move faster than low-pressure areas and tend to overtake them.

As the high-pressure, cold air pushes forward it cuts beneath the low-pressure, warm air. The front between the two air masses is called 'warm' or 'cold' according to the nature of the air mass behind it, and so the arrival of a cold air mass, pushing its way into a warm air mass, is marked by the appearance of a cold front.

The less dense air rides up the edge of the denser air — the front — with only a limited amount of mixing. This causes the rising air to cool, and its water vapour may condense to produce clouds and often precipitation. A cold front slopes back from its direction of movement — think of the cold air mass as dome shaped. As it passes a point on the ground, the cloud and precipitation associated with it begin as the front arrives and continue behind it because, although the point on the ground is then inside the cold air mass, warm air is still being lofted up what is an overhanging slope when seen from below — the inside of the 'dome'. When the whole of the front has

cooling may cause water to condense, but the resulting clouds will be of the stratus, or layer, type. If they produce precipitation, it will tend to be steady rather than in the form of showers. Eventually, warming from the latent heat of condensation may start convection currents that carry air upwards into the overlying air and break down the inversion.

Each type of air mass has its own, distinct characteristics that produce the climate passed, the weather will be that of the cold air mass itself.

After a time, the rear edge of the 'dome' will arrive, as the warm front that marks the boundary between the cold air and the warmer air behind it. This slopes forwards — the inside of the 'dome' again — and the cloud and rain begin some time before the front arrives, and may continue inside the warm air mass, especially if a second cold front is close and advancing. There is a top to the 'dome', where more mixing of air occurs and the fronts become 'occluded' and decay.

Westerly winds prevail in middle latitudes in both hemispheres. Some distance above the surface, they blow parallel to the isobars (they are geostrophic), and are driven partly by the force exerted by the difference in pressure — the pressure gradient — to either side of them. Winds caused by this 'pressure gradient force' are called 'gradient winds'. The temperature gradient, as temperatures generally fall with increasing distance from the Equator, generate a thermal wind. This blows in the same direction as the gradient wind and adds to it. The gradient wind is strongest where the pressure gradient is steepest, and the thermal wind is strongest where the temperature difference to either side is most marked.

The pressure gradient and the thermal gradient are both strongest at a front between a warm, low-pressure air mass and a cold, high-pressure air mass. As the air moves across rough ground it must pass over and around hills, trees, buildings and other obstructions. This causes friction, which reduces the wind speed, and it mixes the air, so that even a clearly marked front is really a region rather than a distinct boundary. The effects of friction and mixing extend some distance above the surface, but their influence decreases steadily with increasing height. The frontal boundary becomes more sharply defined and this, combined with the reduced friction, causes the wind to blow more strongly with increasing altitude.

The jet streams

There are two regions in each hemisphere where major fronts are more or less permanently established. The more northerly is called the Polar Front and marks the boundary between the polar and tropical air masses. The more southerly marks the northern boundary of the subtropical high-pressure zone, at a latitude of about 30°-35°. In these regions, at an altitude between about 30,000 and 40,000 feet (9150-12,200 m), wind speeds in the prevailing westerlies reach their maximum, blowing in fairly narrow bands. These are the 'jet streams'. The more northerly is the Polar Front Jet Stream, the more southerly the Subtropical Jet Stream. Within a jet stream the wind speed can reach 100 to 150 miles per hour (161-241 km/h), and in winter speeds up to 300 miles per hour (483 km/h) have been recorded. The speed generally increases in winter, because it is then that the temperature and pressure gradients are at their steepest — tropical temperatures remain constant, but high-latitude temperatures fall. In summer, the monsoon causes a local reversal in the temperature gradient over Africa and India, with cool air to the south and warmer air to the north, and an easterly jet stream forms temporarily, called the Easterly Tropical Jet Stream. The jet streams influence the development of the monsoon and, more generally, they have a profound effect on the weather.

The jet streams are not continuous around the world, their latitude varies with the seasons and, although (apart from the Easterly Tropical Jet Stream) they blow in an overall westerly direction, they follow a sinuous course. An airline navigator who seeks to exploit a jet stream could find his aircraft being blown strongly to the north or south.

The sinuations are not confined to the jet streams. All the mid-latitude westerlies follow sinuous courses. The mechanism underlying this phenomenon was first explained by the Swedish meteorologist Carl-Gustaf Rossby (1898-1957), and they are known as 'Rossby waves'. The part of a jet stream that approaches most closely to the pole is called a 'ridge', and where it approaches closest to the Equator it is called a 'trough'. Rossby waves result from the interacting effects of the Coriolis 'force' and 'vorticity'. Vorticity is the tendency of a fluid, that is moving in relation to the Earth's surface anywhere except at the Equator, to form an axis and then to spin around it.

The two are distinct and the moving fluid is influenced by their sum, which is called the 'absolute vorticity'. As a body of fluid moves towards the Equator or away

(Opposite) The formation and evolution of a frontal depression. The drawings on the right depict isobars, the boundary between the two air masses (with cold air at the top and warm air below) marked as fronts. The drawings on the right depict the same sequence of events in three dimensions. The arrows show the wind direction. A. The wind flows parallel to the isobars. B. A slight wave develops as cold air begins to undercut the warm air. C. Air begins to flow across the isobars, surrounding what is becoming a distinct region of relatively low pressure, and the continued undercutting produces definite cold and warm fronts at the leading edges of the two air masses. D. As the warm air rises over the advancing cold front, the depression deepens. E. The cold front starts to overtake the warm front, forming an occlusion with the warm air now raised above the cold air, its winds flowing in the same direction as those in the cold air. The system now stabilizes and the depression and fronts disappear.

from it, its absolute vorticity remains constant. The Coriolis 'force', however, increases with distance from the Equator. In the northern hemisphere, if a westerly wind blows a little to the north of its generally eastward track, the Coriolis 'force' acting on it will increase and, therefore, its own 'relative vorticity' will decrease to compensate. The result will be to turn the wind towards the Equator so its direction swings

This sequence of photographs shows the changes in cloud formations as a warm front approaches, preceded by cirrus (opposite, top) and then cirrostratus (opposite, bottom). The nimbostratus (above) behind the front is typical of warm-sector air. It gives way to cumulus in an otherwise clear sky (left), heralding the approach of the cold front.

to the south. This will cause the Coriolis 'force' to decrease, the relative vorticity to increase, and the wind will be swung northwards again.

The Rossby waves are not permanent.

They develop through a cycle over a period of 3 to 8 weeks that ends with the pattern breaking up as the waves form cells. The break-up usually travels from east to west at about 60° of longitude per week. As the

The index cycle. Over periods of three to eight weeks, sinuations in mid-latitude westerly winds and jet streams grow more pronounced until the air movement breaks down into cells, which disappear in their turn as the gently sinuating pattern re-establishes itself.

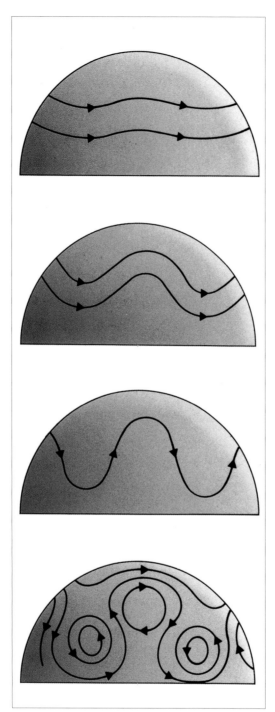

cycle produces increasingly large undulations, cold air on the poleward side extends in tongues towards the Equator and warm air extends towards the poles. This 'index cycle' is most prominent in the Polar Front Jet Stream.

The jet streams occur at major fronts and 'depressions' — local areas of low pressure — develop close to them. The most intense precipitation often occurs directly beneath the jet stream, rather than at the front itself. Inside a jet stream there are regions — 'cores' — where the wind speed reaches a

maximum and air is drawn into the stream, from the poleward side at the entrance to a core and from the equatorward side at its exit. This vertical movement intensifies a depression during its early development.

Depressions move under the influence of the high-altitude westerly winds. This gives them a generally west-to-east direction, but with diversions to either side caused by the Rossby waves. Their speed is approximately 70 per cent of the geostrophic wind speed in the warm air. They sometimes travel quite fast — at up to 20 miles per hour (32 km/h) in summer and 30 miles per hour (48 km/h) in winter.

On a really hot day, the cooling breeze that blows from the sea is one of the delights of a visit to the coast. It is a sea breeze and it provides familiar, tangible evidence of the influence the oceans exert on the climate.

During the day, the land warms rapidly. Warmed air rises, creating an area of generally low pressure. Cooler air from the region of higher pressure over the sea is drawn in to compensate. That is the sea breeze. At the end of the day, as darkness falls, the land cools rapidly, the air over it subsides and becomes denser than the air over the sea. The general cooling of the land also draws subsiding air from higher ground inland down to the coast. The combined effect is to produce a cool breeze that blows from the land to the sea. This is a land breeze.

Land and sea breezes are very local, but almost three-quarters of the Earth's surface is covered by water, and the oceans influence climates on a global scale. Again, there is a well-known and convincing example. Britain and Newfoundland lie close to the coasts of large continents, Britain approximately between latitudes 50 °N and 60 °N, Newfoundland straddling the fiftieth parallel but mainly to its south. Both have humid climates, but they differ markedly in temperature. In London — not the warmest place in Britain — average temperatures range from 40 °F in January to 64 °F in July (4.4-17.7 °C), those of St John's, Newfoundland, from 24 °F in January to 61 °F in August, the warmest month (-4.4 °C-16.1 °C). The difference is significant, especially when you remember that most of Newfoundland lies to the south of London.

Influence of the oceans
Their immense volume and the high specific heat of water make the oceans an

efficient heat store. In summer, the sea-surface temperature is lower than the land-surface temperature, and air crossing from land to sea is cooled. In winter, it is warmed, because the sea-surface temperature is higher than the temperature on land. The air masses over Newfoundland arrive from the North American land mass, those over Britain from the Atlantic Ocean, but this provides only part of the explanation for the climatic difference.

As the air masses arrive, those reaching Newfoundland encounter the cold Labrador Current, bringing water southwards from the Arctic. Those reaching Britain encounter the North Atlantic Drift, a branch of the Gulf Stream that carries warm water northwards from the Equator, via the Gulf of Mexico.

As well as acting as a heat store, the oceans convey heat from the Equator to the polar regions, mainly as a system of currents flowing anticlockwise, called a 'gyre', in every ocean and large sea. Scientists suspect that the Atlantic and, possibly, Pacific gyres are driven, not by the warming of the surface at the Equator, but by the formation of polar sea ice.

When sea water freezes, dissolved salt dissociates from it — melt the ice and you have fresh water. The water adjacent to sea ice contains the salt dissociated from the frozen water and, therefore, it is more saline than water further from the ice. This increases its density. At the same time, it is chilled, and the density of water is at its maximum at 39 °F (4 °C). The dense water sinks, moving away from the edge of the ice and forming a body of deep water that flows towards the Equator. Warmer water is drawn in to replace it and a large-scale circulation establishes itself. Because the flow is anti-clockwise, due to the Earth's rotation, the eastern coasts of the continents lie adjacent to generally cool water and the western coasts to generally warm water. Air reaching western Europe and western North America is warmed just before it arrives.

El Niño
In 1988, and continuing into 1989, North America suffered its worst drought for many years. Crops withered, cattle were slaughtered for lack of feed and large, navigable rivers dried, leaving ships and barges grounded. While North America parched, western Europe had practically no summer

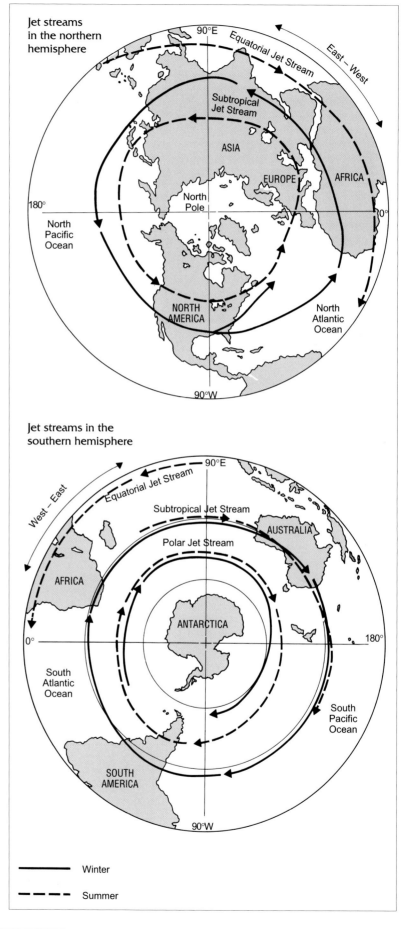

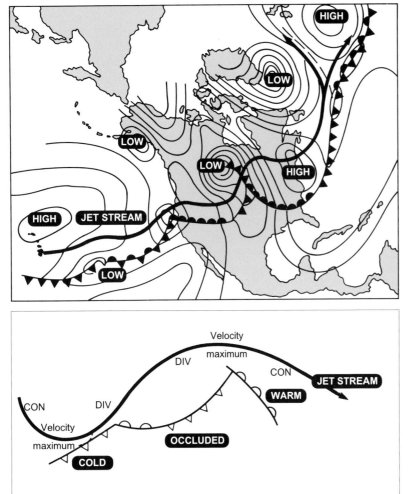

Velocity
maximum
DIV
CON
CON   DIV   **JET STREAM**
                      WARM
Velocity
maximum
OCCLUDED
COLD

Depressions tend to occur close to the regions of maximum jet-stream wind velocity (the 'cores'). Air from the poleward side converges (in the diagram labelled CON) on the jet stream at the entrance to each core and diverges from it (DIV) at the exit; air from the side nearest the Equator diverges from the jet stream at the entrance to the core and converges on it at the exit.

at all; it rained almost incessantly and temperatures were low. Australia, too, had unusually heavy rains in September and, for a time, the desert was carpeted in green. The heavy rains in Bangladesh, southern China and Sudan caused serious flooding. 1988 was a year when much of the world experienced extreme weather. It was not the only year in which unseasonal conditions afflicted the world. The Indian harvest failed in 1987 because of drought, and the winter of 1940-41 was remarkably cold in Britain, with heavy snow, and was followed by a cool, damp summer. For the British, 1965 was another cool, wet year with barely a hint of summer.

These climatic abnormalities puzzled climatologists for a long time, and it is only in recent years that their immediate cause has been discovered. They follow a chain of events, involving the trade winds, that occurs just south of the Equator, in the Pacific Ocean. The events themselves are now fairly well known but, so far, no one knows why they happen. They add up to an

'ENSO' — an 'El Niño-Southern Oscillation'.

'El Niño' is Spanish for 'boy child', or in this case 'Christ child', because the name refers to the fact that the weather associated with an El Niño begins to affect the western coast of South America, especially Peru, in December (mid-summer) — around Christmas. It brings heavy rain — in 1925, another 'El Niño' year, the rains damaged the sun-dried bricks of buildings near Trujillo that had stood for centuries — and a failure of the fisheries on which many coastal communities, and at one time much of the Peruvian national economy, depend. El Niño is an unwelcome gift and the name is used ironically.

On either side of the Equator, the prevailing winds — the trade winds — blow from the east. The trade winds from each hemisphere meet approximately at the Equator itself, at the centre of the low-pressure belt. There the air rises, producing large cumulus (heap) clouds and, beneath them, a region of light winds, or often complete calm. The trade winds are somewhat variable and so is the region where they converge. Draw it on a map and usually it appears as a discontinuous, wavy line. Until weather satellites began to provide meteorologists with a comprehensive picture of conditions day by day, it was difficult to predict where the winds and their convergence would be from one week to the next.

In the days of sail, this was a matter of great importance to mariners. They exploited the trade winds but dreaded the belt between them, which they called 'the doldrums'. In *The Ancient Mariner*, Samuel Taylor Coleridge (1772-1834), the English poet, refers to the plight of a ship ('as idle as a painted ship Upon a painted ocean') trapped in the doldrums, sometimes for weeks on end, with dwindling stores of food and fresh water:

Water, water every where,
And all the boards did shrink;
Water, water every where.
Nor any drop to drink.

Climatologists refer to the region where the trade winds converge as the 'Intertropical Convergence Zone' (ITCZ), although some prefer the 'Intertropical Confluence' (ITC) which, they believe,

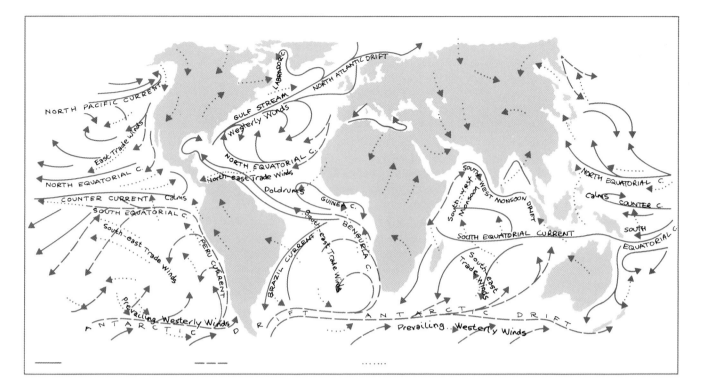

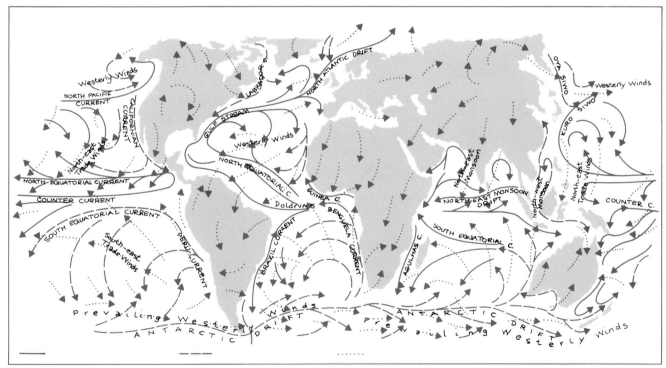

reflects the many and frequent discontinuities in the convergence. The ITCZ or ITC is an average condition, not one that is fixed and permanent.

The great ocean gyres are believed to be driven mainly by the formation of sea ice but, at the surface, ocean currents generally are also driven by the wind. Where its direction is variable, the wind pushes the water first one way and then another, so the overall effect is only to influence the speed and force of existing currents and tides. Where the wind direction is more reliable, however, the effect is greater. The trade winds, blowing from north-east to south-west and south-east to north-west, produce an Equatorial Current in each hemisphere. The water carried by the Equatorial Currents is replaced by the Equatorial Countercurrents, which flow in the opposite

Global wind patterns and their relationship to the Equatorial Currents and Counter Currents.

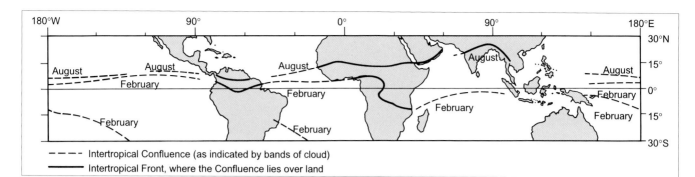

- - - - - Intertropical Confluence (as indicated by bands of cloud)
———— Intertropical Front, where the Confluence lies over land

direction at a much deeper level.

South of the Equator, the Equatorial Current moves surface water away from South America and towards Asia. The surface water is warmed strongly by the Sun and, therefore, the effect is to move warm water westwards. It accumulates near Indonesia, where the depth of warm surface water is much greater than it is near the western coast of South America.

A cold current, called the Peru (or Humboldt) Current, moving deep below the surface, carries Antarctic water northwards through the eastern South Pacific as part of the ocean gyre. Ocean water forms layers, rather like the layers of the atmosphere, in which cold, dense water lies beneath warm, less dense water. A cold current, moving towards the Equator, flows deep below the surface but, near its edges where moving water is adjacent to still water, or where its flow is perturbed, there are upwellings. As it enters the tropics, the upwellings in the Peru Current come close to the surface because there is only a shallow layer of warm water above them. The water of the Current is rich in dissolved oxygen and nutrients and, therefore, its upwellings support abundant marine life — including the vast populations of sea birds that nest on rocky islands which they have covered with the thick layers of droppings local industries market as 'guano', a rich fertilizer.

From time to time, the difference in temperature between them causes air to flow from the high-pressure area in the south-eastern Pacific into the low-pressure area around Indonesia. This is associated with a southward displacement of the ITCZ (or ITC) — a 'Southern Oscillation' — which amounts to a southward shift of the entire system of atmospheric pressures and winds.

The doldrums now occupy the region just south of the Equator. The trade winds slacken, the Equatorial Countercurrent strengthens and the Equatorial Current

weakens or even reverses its direction. Instead of being driven westwards, warm water accumulates in the eastern Pacific. The upwellings of the Peru Current are suppressed by the greater depth of warm surface water, leading to a collapse of the fisheries (although there is usually an influx of tropical fish, such as yellowfin tuna).

The climatic consequences of this disturbance to the usual patterns are felt throughout the world — and a similar, though smaller, El Niño may sometimes occur in the Atlantic. Indeed, the link is now so well established that it can be used to make general predictions of the kind of weather people may anticipate for months ahead.

El Niño has an opposite, known as an anti-El Niño or 'La Niña, in which the Equatorial Current flows more strongly and the trade winds intensify. As you might expect, this also produces climatic extremes, but in the other direction. Where an El Niño — or, more correctly, an ENSO event — produces drought, for example, its opposite is associated with heavy rain. No one knows what triggers these chains of events, but some scientists suspect they may be occurring more frequently. If so, this may be related to overall global climatic changes.

(Opposite) The prevailing winds push the surface waters of the sea to form currents.
(Above) The location of the Intertropical Confluence.

Cross-section of the Pacific, just south of the Equator, showing the layers of warm and cool water. Prevailing (trade) winds move warm surface waters in a westerly direction, as the South Equatorial Current, so the depth of warm water is much greater near Indonesia than it is off the South American coast. The movement of surface water is compensated for by a flow of cool water in the opposite direction at greater depth.

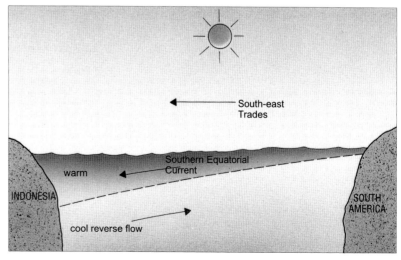

THE CLIMATES OF EARTH

There is an intimate and reciprocal relationship between air masses and the large surface areas over which they form and acquire their characteristics. The weather we experience is brought to us by a particular air mass but, as soon as it reaches us, it is modified by the very local circumstances it encounters. Those circumstances — such as the latitude, the presence of land or water and the surface topography — alter the air mass. It may be warmed or cooled, gain or lose moisture, and so its qualities are constantly changing as it moves, and the mass which passes on is subtly different from the mass that arrived.

Principal types of air masses
Air masses deliver weather but they are also products of the climatic conditions that formed them. This means that the labelling of an air mass, as tropical, arctic or polar, continental or maritime, provides a broad description of the climate of its source region. When we are visited by continental tropical (cT) air, for example, the label tells us something of the kind of weather we may expect from it. The label also tells us that this weather will be our own local variation of something like the climate that is typical of the interior of a continent in a low latitude. Clearly, the conditions under which air masses form provide the 'raw material' for the study of the world's climates.

The name 'continental tropical' (cT) is slightly misleading. These air masses form in the interior of continents in low latitudes — but in the subtropics, not the geographical tropics. Tropical (that is, Equatorial) air is more or less confined to a belt either side of the Equator bounded by the Tropics of Cancer (northern hemisphere) and Capricorn (southern hemisphere). This is the region where air movement is mainly vertical and dominated by the Hadley cells.

In the northern-hemisphere winter, the northern boundary of cT air extends from a little way inland from the Atlantic coast of Africa, along the southern shore of the Mediterranean, north into eastern Turkey, across the Caspian Sea and eastwards across central Asia to a position approximately north of Bangladesh. The southern boundary is where cT air meets Equatorial air. It crosses Africa, passing just north of the West African coastal belt and turns north, excluding the Horn of Africa, follows the southern coast of the Arabian Peninsula, and crosses into Asia to the north of Afghanistan, Pakistan and India, to end about at the north-eastern corner of Bangladesh. Another cT region occurs in the western and south-western United States and Mexico, extending from around Portland, Oregon, down the western side of the Rocky Mountains and into Mexico as far as the Tropic of Cancer. In the southern hemisphere winter, cT air covers most of southern Africa and the interior of Australia.

In summer, the cT air masses expand polewards over Africa and Eurasia. They cover most of Spain (but not Portugal), and extend across central Europe and Asia as far as the southern shore of the Sea of Okhotsk. In North America, however, the cT area does not change and in the southern-hemisphere it contracts.

Aridity is the outstanding characteristic of continental tropical air. The regions in which it forms include the Sahara, Arabian, Gobi, Australian and Kalahari Deserts, as well as the deserts of North America and Mexico. A desert is likely to occur wherever the rate of evaporation exceeds the rate of precipitation over a prolonged period. This sounds precise but, in practice, it is not so simple. A region may receive all of its rain during one season, for example. During the rainy season, rainfall exceeds the rate of precipitation comfortably, but for the rest of the year — which may be the remaining nine or ten months — very little rain falls. The 'precise' definition would suggest that such a region is sometimes desert and sometimes not. Over most of the world, however, an average annual precipitation of less than 10 inches (254 mm) is almost certain to produce a desert.

The air comprising the mass reaches the subtropics as the descending side of the Hadley cell circulation and this explains its aridity. Virtually all of the water which evaporated into it in Equatorial regions condensed and fell as rain, inside the Equatorial belt — the humid tropics — as the air was lofted all the way to the tropopause. By the time it descends it is extremely dry. The inhabitants of Ain Salah, for example, in central Algeria, can expect about one very heavy shower of rain every 10 years. On average, Cairo has about 1.1 inches (28 mm) of rain a year and Baghdad has about 5.5 inches (140 mm). Rainfall is a little higher along the coasts. Jedda, on the Saudi Arabian coast of the Red Sea, not far from Mecca, receives an average of 2.5 inches (63.5 mm).

The central Asian deserts receive 2 inches (50 mm) of rain a year or less, but conditions are much less severe close to their northern borders. With about 8 inches (203 mm) of rain a year, the grasslands around Ulan Bator, Mongolia, although they are patchy, support livestock husbandry.

The North American desert is almost as dry as the Sahara. At Yuma, Arizona, for example, close to the border between the United States and Mexico, the average annual rainfall is about 3.4 inches (86 mm). Conditions in Australia are not quite so extreme. Alice Springs, for example, about at the geographical centre of the country, has an average of 9.9 inches (251.5 mm) of rain a year, and Kalgoorlie has about the same. Nevertheless, more than one-third of the whole area of Australia — about one million square miles (2.6 million sq km) — receives less than 10 inches (254 mm) a year. The arid lands of southern Africa, however, are very dry indeed. Walvis Bay, Namibia, has only about 0.9 inches (23 mm) of rain a year, one-third of it falling in March.

Faced with rainfall averages experienced in parts of the world where the climate is unlike those which are familiar to people living in temperate regions, there is a risk of comparing one with another and forgetting just how low all the figures are. For comparison, therefore, the average annual rainfall in New York is 43 inches (1092 mm) and in London it is 24 inches (610 mm) — and neither of these cities has what a climatologist would describe as a humid climate.

Such rain as does fall tends to do so in a distinct rainy season. In the north of Australia, the rain falls mainly in summer — due partly to the Asian monsoon — and in the south it falls in winter. The east has rain all year round, but the west and the centre of the country have no reliable rainy season. The Sahara is more likely to have rain in summer but further east, in Asia, rain tends to occur in winter.

Extreme temperatures are associated with the aridity. Because the rainfall is very low, for most of the time the sky is clear and the ground dry. Clouds do cross the sky sometimes but, even when rain falls, it usually evaporates again before reaching the ground. Large areas of the surface consist of sand, gravel or larger fragments of rock, with a relatively low specific heat, which warm and cool rapidly to produce high summer daytime temperatures. The surface rocks are often too hot to touch, and the temperature of the surface sand has been known to rise to more than 170 °F (77 °C). Below the surface, however, the sand is much cooler and desert animals find tolerable conditions by burrowing into it. The air temperature often exceeds 100 °F (38 °C) and on 13 September 1922, Azizia, about 25 miles (40 km) south of Tripoli, experienced the highest temperature ever recorded on Earth — 136 °F (57.8 °C). The temperature falls rapidly after sunset, by as much as 30 °F to 40 °F (17-22 °C) to give night-time temperatures of around 60 °F to 70 °F (15.5-21 °C).

Winter temperatures are much lower, although they are still high compared with those of temperate regions. Generally, daytime temperatures seldom fall much below about 60 °F (15.5 °C). The nights can be cold, however, with the temperature falling to about freezing.

Water is a valuable resource throughout the interior of continents but, for the people who live there, cT air does not bring an unhealthy climate or even, necessarily, an unpleasant one. The dryness of the air allows the human body to cool itself effectively by perspiring — humid heat is much more uncomfortable, and unhealthy, than dry heat. It is cT air, after all, which gives southern Europe its warm, dry summers. Sea-level temperatures are high, but continental interiors are often mountainous and at higher elevations conditions are often very pleasant, because of the extent to which air cools with increasing altitude. Tehran, in Iran, for example, is 4000 feet (1220 m) above sea level and its annual temperatures range from about 36 °F (2.2 °C) in January to about 85 °F (29 °C) in July, with an annual average of a comfortable 62 °F (16.6 °C).

On the poleward side of the subtropical

belt of cT air, the continental air is polar — cP. No cP air masses form in the southern hemisphere but, in the northern hemisphere, they are extensive. In summer, the southern boundary of cP air in North America lies almost exactly along the Canadian border and, in winter, it expands southwards in a great arc, the southernmost point of which is in the region of Dallas, Texas. In the Old World, the coastline marks the northern boundary of cP air in summer and in winter. In the west, the summer boundary follows a line approximately from Murmansk, through Leningrad, to somewhere near Smolensk, then eastwards across the middle of the continent all the way to the Sea of Okhotsk. In winter, the western boundary lies further west, roughly along the border between Sweden and Norway, and the southern boundary lies much further to the south.

The air is continental and this means it is relatively dry, although the annual precipitation is much higher than in the cT regions to the south. In the Canadian prairies, for example, Medicine Hat, Alberta, has an average of 13.6 inches (345 mm) a year distributed fairly evenly. This may not sound like a dry climate but in the wettest month, June, the average rainfall is only 2.3 inches (58 mm) and many months have 1 inch (25 mm) or less. Winnipeg, Manitoba, has an average 19.7 inches (500 mm) a year. These figures are typical for regions covered by the North American cP air mass.

In Eurasia, the annual precipitation at Tomsk averages 19.9 inches (505.5 mm), again with June, as the wettest month, receiving 2.7 inches (68.6 mm). Moscow is somewhat wetter, with an average 24.8 inches (630 mm) a year. The figures, which are also typical for the Eurasian cP area, are similar to those of North America. Continental polar air is fairly dry, but nowhere is it nearly dry enough to produce desert conditions.

Much of the precipitation occurs as snow, because, as the 'polar' in its name suggests, cP air is cold in winter. In the Canadian prairie provinces, mid-winter temperatures of -15 °F (-26 °C) are common, -30 °F (-34 °C) is reached once or twice in most winters and, in a really hard winter, the temperature can go down to -40 °F (-40 °C) or even lower. These low temperatures are due partly to the elevation of the land which increases fairly steadily as you travel westwards across the continent — although Winnipeg, where the average January temperature is 1 °F (-17 °C) is only 786 feet (240 m) above sea level. Medicine Hat, on the other hand, is at 2365 feet (721 m) and Calgary, Alberta, is at 3540 feet (1080 m) — yet both towns have higher average January temperatures than Winnipeg (14 °F, -10 °C and 16 °F, -9 °C respectively). Rivers and lakes are covered with ice for the whole of the winter, and the soil moisture also freezes to a depth of several feet.

Low though they are, these temperatures are by no means the lowest that have been recorded. Poplar River, Montana, has experienced -63 °F (-53 °C) and at Winnipeg, Medicine Hat and Edmonton, Alberta, it can be colder than -50 °F (-45.5 °C). The averages close to the Rocky Mountains (in Calgary, for example) are raised because of the frequent chinook winds. Eurasian winter temperatures are very similar. The average January temperature in Tomsk is -6 °F (-21 °C) and in Moscow 16 °F (-9 °C). As in North America, however, the average embraces a wide range. In January, the temperature in Moscow can fall to -44 °F (-42 °C) and in Tomsk to -58 °F (-50 °C).

Föhn winds

The chinook is a wind of the 'föhn' type, which also occurs in parts of Europe and central Asia. Such winds occur when there is a temperature inversion at about the height of the summits of a mountain range. Air moving against the mountains is forced to rise, but its upward movement is checked by the inversion — the air cannot rise through the less dense air above it. The air crosses the mountains, then spills down the leeward side, warming adiabatically as it descends and crossing the lower ground as a warm, dry wind. The effect can be spectacular. In Calgary, a chinook can raise the temperature by 60 °F (33.3 °C) in an hour and, on one occasion, in January, 1966, a chinook reaching Pincher Creek, Alberta, raised the temperature by 38 °F (21 °C) in about four minutes. In Tashkent, the temperature has been known to rise from about freezing to more than 70 °F (21 °C) during a föhn. The arrival of the wind is easily predicted. A distinctive arc of blue sky appears beneath the cloud base over the mountains, and the time that elapses between the appearance of the arc and the arrival of the wind depends on the distance between the observer and the mountains that the warm air has to traverse.

Chinooks and föhn winds help to raise average winter temperatures and often come as a great relief to people enduring extreme cold, but they can also cause problems. In the mountains, the sudden rise in temperature can thaw snow sufficiently to cause avalanches. At lower levels, where the wind can move back and forth across an area, causing the temperature to rise and fall rapidly and repeatedly, snow will melt and then freeze again, forming a substantial layer of ice on roads.

Continental polar air masses

If 'polar' suggests 'cold', 'continental' must suggest 'extreme' and although cP air masses are extremely cold in winter, they are warm in summer. It is not unknown for the people of Moscow to bask at 99 °F (37 °C) in July, and those living in Tomsk sometimes enjoy much the same degree of warmth although, in both cases, these are about 30 °F (17 °C) higher than the July average. North America is not quite so hot in summer, but 80 °F (27 °C) is not unknown in Winnipeg, and in Medicine Hat the July temperature can 85 °F (29 °C).

In winter and in summer, the usually low humidity brings blue skies and sunshine, so that both heat and cold feel less extreme than a visitor from a more humid climate might expect. Indeed, strangers can be tempted to go outdoors in winter without gloves or ear coverings quite unaware of the serious risk of frostbite — which can occur quickly and without the victim feeling it.

At ground level there is little difference between cP air and the air mass that surrounds the poles themselves — continental arctic (cA) air. The difference becomes apparent only in the middle and upper levels of the troposphere, above about 20,000 feet (6100 m). At this height, cA air is the colder of the two, a fact that is important in the mechanisms which are involved in the seasonal depletion of the ozone layer in Antarctica and, in some years but more briefly and to a lesser extent, in the Arctic (see page 161).

Continental arctic air masses

Continental arctic air masses form only around the North and South Poles, and in

the northern hemisphere the cA air mass disappears entirely in summer, to be replaced by maritime air. In the southern hemisphere, however, cA air covers the whole of Antarctica throughout the year, its boundary following the coastline. The coastline is not visible because the land is surrounded by sea ice with no clear shoreline where land and sea meet. Ice that extends from the coast over the sea is called an 'ice shelf' and, in the Ross Sea, the shelf is about 175 feet (53.4 m) thick. In most places, the shelves extend some 200 to 300 miles (322-483 km) from the edge of the land, advancing in winter and retreating in summer and beyond them the sea is covered with ice floes — looser masses of ice — and icebergs (literally, 'ice mountains'), which periodically break away from the main ice mass and drift into higher latitudes.

The entire land surface of Antarctica lies hidden beneath a sheet of ice that contains nine-tenths of all the ice on Earth. Its thickness varies from place to place but, on average, the sheet is nearly 7000 feet (2135 m) thick and its total volume has been estimated as about 7 million cubic miles (30 million cu km). In the (extremely unlikely) event that all of this ice were to melt, sea levels throughout the world would rise by some 250 feet (76 m). Beneath the ice, Antarctica is the fifth largest continent and the highest — over most of it the ground level is about 8500 feet (2.6 km) above sea level.

Any description of Antarctica is richly embellished with superlatives. It is, indeed, a place of extremes of every kind. It is also a place of contrasts and, now and then, of surprises. Not all of the land is buried beneath ice, for example. There are 'dry valleys' which are free from ice and where lichens grow, exploiting the brief period in summer when they have access to sunshine and a little liquid water.

The air mass is continental and, like other continental air masses, its pressure is high and the air is dry. Technically speaking, most of Antarctica is a dry desert and the aridity of the interior is comparable to that of the Sahara. This seems paradoxical, given the vast quantity of frozen water that lies on the surface and the frequency of blizzards, but it is true, nevertheless.

Blizzards develop quickly. Within a few hours completely calm air, with barely a breath of wind, can give way to a howling

Icebergs, or 'ice mountains' as the word means literally, periodically break away from the main ice mass and drift into higher latitudes. The part that you can see really is the 'tip of the iceberg' for almost 90 per cent of it is under water.

gale, usually from the south, in which wind speeds can reach 70 miles per hour (113 km/h) or more and visibility is reduced to zero in a 'white-out', when snow is driven almost horizontally against the white background. This is not precipitation, however. A 'blizzard' is strong wind, not a heavy snowfall, and the snow it carries is blown up from the surface. It is the cold-climate equivalent of a desert sandstorm, and no one supposes that the wind-driven

sand is falling, like rain, from the sky.

True blizzards occur only in very cold conditions, when snow takes the form of very small crystals, almost as fine as dust, lying loosely on the surface. Nor does the thickness of the ice sheet prove that snow falls heavily or even regularly. Except around the coasts in summer, the temperature in Antarctica never rises above freezing, and in the interior all precipitation falls as snow. Some is lost from the surface rapidly, by melting in the sunshine and rapid evaporation into the dry air, or directly, by sublimation — the alteration of a solid (ice) into a gas (water vapour) without passing through the liquid phase. The snow that survives evaporation or sublimation remains on the surface and accumulates. The ice sheet is the product of a very long period of accumulation. Some scientists believe that the Antarctic climate was once less arid than it is today and that the present ice accumulated during an earlier interglacial (see page 125). If this is so, the thickness of the ice sheet is quite unrelated to today's rates of precipitation.

It is more difficult to measure snowfall than rainfall, especially in Antarctica, because the snow tends to be driven across the measuring gauge rather than falling into it and, to complicate matters further, old snow, blown up from the ground, is mixed with the fresh snow that is falling. When the amount of fallen snow has been measured the figure must then be converted into a 'rainfall equivalent', because snow is made from loose crystals separated by air spaces, so that much of the thickness is accounted for by air. The meteorologist has to sample the snow, measure its thickness, melt it, and see how much liquid water it represents when it is just warm enough to remain liquid — the temperature is important, because water expands as it warms. Having done that, an allowance must be made for the amount that is lost by evaporation and sublimation.

With all these qualifications, it is possible to give estimates of the amount of precipitation and the amount remaining after evaporation and sublimation, converted into 'rainfall equivalent'. Between latitudes 75 °S and 80 °S, the average annual precipitation is about 3.2 inches (81 mm) of which about 1.6 inches (41 mm) remains on the surface. Near the Pole itself, from latitudes 85 °S to 90 °S, the same amount remains lying, but only because a smaller amount is lost. The actual precipitation is less, averaging about 2.8 inches (71 mm). This is about the same as the annual rainfall at Touggourt, in northern Algeria.

Perhaps Antarctica would have to fight off competitors to establish its claim to be the driest desert on Earth — though it might well win the contest — but there can be no doubt that it is by far the coldest. It is always cold, but, for geographical reasons,

the seasonal variation in temperature is much larger in Antarctica than it is in other parts of the world.

The seasons are caused by the Earth's orbit around the Sun, and the fact that the planet's axis of rotation is slightly tilted so that first the northern hemisphere is tilted towards the Sun and then the southern hemisphere. This has no effect at the Equator which faces the Sun at the same angle throughout the year but, in higher latitudes, the effect increasing with distance from the Equator, it increases the number of hours of daylight in summer and decreases it in winter. The effect reaches its maximum near the poles. The Arctic and Antarctic Circles are defined as the limits, polewards of which there is at least one period of 24 hours during the year when the Sun does not appear above the horizon, and one 24-hour period when it does not sink below the horizon. In Antarctica, all of which lies inside the Antarctic Circle, this means there is a period without darkness during the summer and a period during the winter when there is no daylight. In winter, when there is little or no direct sunlight, temperatures fall very low but, in summer, when sunshine is almost continuous, they rise.

How air mass and climate interact
The climate affects the air mass and the air mass also affects the climate. Antarctica is almost completely covered by snow, and the air above it is mainly dry and free from cloud — despite the dark days of winter, there are often more hours of bright sunshine during the year than in even the sunniest place in Britain. This has two consequences. In summer, the brilliantly white surface reflects back into space some 80 to 90 per cent of the radiation falling on it (this is its 'albedo', described more fully on page 138). Were the ground a darker colour, it would absorb more radiation and become warmer. The radiation that is absorbed penetrates the snow to a depth of no more than about 4 feet (1.2 m), so its warming effect is limited. That is the first consequence, of the nature of the surface. The second, due to the air itself, is that the little heat which is absorbed is readily lost. The sky is clear and dry. When the ground — or any other body — has absorbed radiation it begins to re-radiate it at longer wavelengths. During daylight, the rate of absorption exceeds the rate of re-radiation,

and the ground warms. At night, however, when there is no incoming sunlight, re-radiation becomes dominant and the ground loses heat by its own radiation. This long-wave radiation can be intercepted in the atmosphere by water droplets and molecules of water vapour — the 'greenhouse effect' (see page 161). Such interception warms the molecules, which also re-radiate, and the result is to reduce the rate at which energy — heat — is lost from the ground. When the air is dry and free from clouds, however, the 'blanket' is removed. It is why, in our own winters, clear nights often lead to frost, and cloudy nights are warmer. In Antarctica the effect is more extreme and the ground cools until its temperature has fallen very low indeed.

At the onset of winter, temperatures fall sharply owing to radiative cooling. Then they rise a little because the intensely cold air produces a change in the atmospheric circulation that draws in warmer air from lower latitudes and brings cloud which reduces the radiative loss. Towards the end of winter, this circulation ceases, radiative cooling becomes dominant again and, shortly before the return of the Sun, there is a second period of very low temperature.

Inland, where the air mass forms, summer temperatures are usually between about -30 °F and -4 °F (-34 and 20 °C). In winter they fall to between -40 °F and -94 °F (-40 °C and 70 °C). These are average temperatures and it can be much colder, especially in the mountains. A mountain chain effectively divides the continent into two unequal parts: in East Antarctica (the side facing New Zealand) there is an extensive high plateau, at more than 13,000 feet (4000 m) where the average winter temperature is -69 °F (-56 °C) and sometimes it falls to -125 °F (-87 °C).

There is no land at the North Pole although, in winter sea ice covers the entire area between Asia and North America, extending to the southern tip of the Bering Peninsula and, in the western Atlantic, to the southern tip of Greenland. In the eastern Atlantic the warm North Atlantic Drift prevents the ice sheet from covering the area west of the Kola Peninsula, so the North Cape, on the northern shore of Norway, is ice-free throughout the year. In winter, a cA air mass forms over the whole of the ice-covered area. It disappears in the summer, and so the only climatic conditions

associated with it develop in winter.

Arctic winters are cold, but not so severe as those of Antarctica. At Coppermine, for example, in the far north-west of Canada not far from the Alaskan border, February is the coldest month with an average temperature of -20 °F (-29 °C). Coppermine is close to sea level, but most of this region is low lying and so its climate is probably typical. Winter temperatures sometimes fall to below -50 °F (-45.5 °C) and, cold though this is, it might be considered mild in the interior of Antarctica. In the far north of Siberia, temperatures fall to about -58 °F (-50 °C). Verkhoyansk, the northern hemisphere 'cold pole', lies just inside the Arctic Circle, to the east of the Verkhoyansk range of mountains, but in a valley — a frost hollow. The temperature sometimes falls to -68 °F (-55.5 °C), but in East Antarctica such a temperature — the lowest recorded in the northern hemisphere — is merely average for the time of year.

The area covered by the northern cA air mass is dry, but not so extremely arid as Antarctica. The Kola Peninsula, for example, receives about 16 inches (406 mm) of precipitation a year, and Coppermine records an average of about 11 inches (279 mm), although some of the northern-most islands off northern Canada have only about 3 inches (76 mm) a year.

Continental air masses are dry, and warm if they are tropical and cold if they are polar or arctic. The seasonal extremes that are typical of climates in the interior of North America and Eurasia occur because these regions are dominated by polar air masses in winter and in summer by tropical air masses, or by transitional air masses that lie between polar and tropical air and have properties intermediate between the two.

Maritime air masses

Maritime air masses are moist. Maritime tropical (mT) air masses form in low latitudes in the Atlantic and Pacific Oceans. Geographically, they are the most extensive of all the air masses — about half of the southern hemisphere is covered by mT air. In the Indian Ocean, and in the Pacific around Indonesia and Papua New Guinea, monsoon air forms between the Equatorial air and the cT air mass on the poleward side — in central Asia and Australia.

Like cT air masses, mT air consists of Equatorial air on the descending side of the

Hadley cells and it forms large areas of high pressure. In both cases, the descending air is dry and warms as it descends. Maritime air, however, descends to the ocean surface and water evaporates into it which increases its humidity in the lower levels — although it remains dry aloft. Because the air is descending, it is inherently stable and as it moves from lower to higher latitudes its water vapour tends to condense into stratiform (layered) clouds. The air is mild or very warm throughout the year, often with a small seasonal variation in temperature compared with that of cT air, due partly to the moderating influence of the ocean and partly to the cloud, which has a cooling effect in summer. In the southern hemisphere, there is no significant change in average temperatures between the Equator and about 40 °S. Near sea level in the Azores, which lie in the northern hemisphere mT air mass, the average temperature is 48 °F (9 °C) in January and 82 °F (28 °C) in July. This seasonal difference of 34 °F (19 °C) seems large but it is less than, for example, that in Kuwait, where the average temperature is 55 °F (12.7 °C) in January and 95 °F (35 °C) in August, a difference of 40 °F (22 °C).

Madagascar (the Malagasy Republic) lies within the mT air mass of the western Indian Ocean. It is windy, especially in the north, because it lies within the trade-wind belt. The ridge of mountains that runs the length of the island deflects and slows the trade winds over the southerly regions. The rainfall occurs mainly in summer. The central region has an average of 40 inches to 80 inches (1016-2032 mm) a year, but more than 100 inches (2540 mm) fall in some of the hilly parts of the east. The north is drier, with only about 38 inches (965 mm) and parts of the south-west and southern tip are drier still, with 20 inches (508 mm) a year or less. Storms are common and, in the northern part of the island, thunderstorms are very frequent. The climate is warm, with summer temperatures averaging about 80 °F (27 °C) near sea level and about 70 °F (21 °C) on high ground at about 4500 feet (1370 m). Winter temperatures range from below 60 °F (15.5 °C) on high ground to about 65 °F (18 °C), so the annual temperature range in different parts of the island is between 7 °F and 15 °F (4-8 °C).

The islands of the Caribbean also lie within mT air throughout the year but, in summer, Equatorial air moves north, the subtropical high-pressure area moves away to the north-east and the islands are affected by fronts between the subtropical high and the Equatorial low. Summer is the rainy season, although some rain falls in winter. The rainfall varies according to the topography. Kingston, Jamaica, has an average of only 31 inches (787 mm) a year, but places in the Blue Mountains have more than 200 inches (5080 mm). The summer rains often occur as afternoon storms. Winter rains are lighter but more persistent.

As well as lying within a source region for mT air, the West Indies lie close to the source of the Gulf Stream, and the sea-surface temperature ranges from a little below 60 °F (15.5 °C) to over 80 °F (27 °C) or even over 90 °F (32 °C). Islands lying within mT air enjoy an apparently idyllic climate, but it suffers from a serious disadvantage — hurricanes are common. Those which cause severe damage in the south-eastern United States, and occasionally cross the Atlantic to affect western Europe, originate in the Gulf of Mexico and Caribbean. 'Hurricanes' are known as 'tropical cyclones' in the Indian Ocean (and as 'typhoons' in eastern Asia), and Madagascar also suffers from them.

How hurricanes develop

Hurricanes seem to develop when the Equatorial low-pressure area moves some distance from the Equator. The sea-surface temperature must be high — more than 80 °F (27 °C) — over a large area that covers a region far enough from the Equator for the Coriolis 'force' to be significant (it is zero at the Equator), but not so far from the Equator that it lies beneath the subtropical jet stream. There must also be an area of high atmospheric pressure, with air flowing from it, close to the tropopause. Most hurricanes begin 5-10° from the Equator in the 'doldrums' area.

The combination of a very warm sea surface, calm winds and high pressure aloft causes strong convection and the development of a local area, up to about 400 miles (644 km) in diameter (although they are often larger in the China Sea), of intensely low pressure. Warm, very moist air rises rapidly and, because of the Coriolis 'force', air drawn in at low level to equalize the pressure is deflected, so the whole system twists about its vertical axis, forming a

In Madagascar, the rain falls mainly in summer and is usually intense, with frequent thunderstorms.

(Opposite, top) Canada has a climate produced by cP air, relatively dry with warm summers and cold winters. This is the Yukon. (Opposite, bottom) Britain, although in a similar latitude, has a mild maritime climate, typical of mP air. This is the Vale of Evesham, an important fruit-growing area.

vortex. The rising air cools, but the cooling is offset by the latent heat of condensation as its water vapour condenses to form massive cumulus and cumulonimbus clouds, towering to a height of 40,000 feet (12.2 km) or more. The system 'feeds on itself', producing a vortex of rising air that is constantly warmed. The rising air increases the high-altitude pressure, increasing the rate at which air flows out from it, and this, in turn, intensifies the low pressure at the surface. Air is drawn downwards at the centre, from high pressure above towards low pressure below, spills to the sides and is warmed at the surface and then rises again as the vortex of intense winds.

The 'eye', at the centre, has a diameter of some 200 to 300 miles (322-483 km). Within it, the skies are clear and there is

little or no wind. In the vortex, however, the wind speed is never less than 75 miles per hour (120.7 km/h) — by definition, the minimum speed for a wind classified as 'hurricane force' — and they often reach 120 miles per hour (193 km/h). The most severe winds usually last for only a few days, after which the conditions producing the hurricane begin to break down, and the hurricane system moves as a whole, usually at up to about 15 miles per hour (24 km/h), the precise speed and direction depending on the movement of the warm upper air. When a hurricane crosses land, it causes extensive damage and, of course, it presents a major hazard to shipping while it is at sea. Even after it has crossed the Atlantic all the way to Europe, a particularly vigorous hurricane can bring storm-force winds that

bring down trees and damage buildings.

Maritime polar (mP) air is associated with regions of low atmospheric pressure and it brings cool, wet weather in both summer and winter. The climate of Tierra del Fuego, which forms the southern tip of South America, is produced by mP air throughout the year and is fairly typical. In winter, temperatures fall to about freezing and, although the temperature rarely remains below freezing for very long, Ushuaia, Argentina, is prone to brief frosts in any month. In summer, the average temperature is around 50 °F (10 °C), although it can rise as high as 80 °F (27 °C). The winter is slightly wetter than the summer, but the difference is not marked and rainfall is distributed fairly evenly through the year. At Ushuaia, and at nearby Punta Arenas, Chile, the average annual rainfall is about 20 inches (508 mm).

The Aleutian Islands, in approximately the same northern latitude as the southern latitude of Tierra del Fuego, also have a climate typical of mP air. Because they form part of Alaska, we may be inclined to think of these islands as having an Arctic climate. In fact, they lie in about the same latitude as the British Isles — and about half of Alaska itself lies to the south of the Arctic Circle. Temperatures in the Aleutians vary little between summer, when the average is about 50 °F (10 °C), and winter, with an average just a degree or two above freezing. The yearly average of about 38 °F (3.3 °C) is high enough to give a growing season for vegetables of 135 days, lasting from May to September, and there are many flowering herbs, although the drying effect of the almost constant winds mean there are few trees — the offshore islands of northern Europe have few trees for the same reason. The rainfall is heavy — about 80 inches (2000 mm) a year — and fogs are almost perpetual.

A small maritime arctic (mA) air mass forms in winter over the Ross Sea, Antarctica, and a much larger one covers the polar regions of the northern hemisphere in summer, when continental arctic air flows across open water. This happens during the Antarctic winter as the extremely dense air over the continent spills outwards, eventually to the open sea. In the northern hemisphere summer, some of the sea ice melts, as does some of the snow and ice over land, and cA air is affected by it. Continental polar air is similar to cA air, but moister.

About four-fifths of the total area of Greenland, which lies inside cA air in winter, but mA air in summer, is covered by an ice sheet with an average thickness of 5000 feet (1.5 km) — and twice that thickness in places — but away from the ice sheet temperatures are high enough to support vegetation, and local people are able to grow vegetables. The name, 'Greenland', is said to have been bestowed on this vast island in the tenth century by Eric the Red in order to encourage settlers to move there from Norway. His venture was not particularly successful but, although the climate is harsh, it is not quite so harsh as its ice sheet might suggest. Snow may fall at any time of year and at Angmagssalik, on the east coast, just south of the Arctic Circle, the average temperatures range from about 19 °F (-7 °C) in January to about 45 °F (7 °C) in July and

(Opposite, top) North Africa lies within the subtropical arid zone, dominated by cT air. This is the Sahara Desert. (Opposite, bottom) The climate of the Caribbean is produced by mT air and is considered by many people to be idyllic. This is Martinique.

The scale was devised in 1806 by Admiral Sir Francis Beaufort (1774-1857) in relation to the amount of canvas a sailing ship could carry. The scale was adopted by the British Admiralty in 1838 and by the International Meteorological Committee in 1874.

The Beaufort Wind Scale

Force	Velocity mph (km/h)	Description	Effect
0	less than 1 (1)	Calm	Smoke rises vertically
1	1–3 (1.6–4.8)	Light air	Smoke shows direction of the wind but wind vanes do not move
2	4–7 (6.4–11.3)	Light breeze	Leaves rustle; wind vanes move; wind felt on face
3	8–12 (12.9–19.3)	Gentle breeze	Leaves and twigs move constantly; light flags move
4	13–18 (20.9–29.0)	Moderate breeze	Dust and loose paper blow about; small branches move
5	19–24 (30.6–38.6)	Fresh breeze	Small trees with leaves sway
6	25–31 (40.2–49.9)	Strong breeze	Large branches move; difficult to use umbrellas
7	32–38 (51.5–61.1)	Moderate gale	Whole trees move; pressure when walking into the wind
8	39–46 (62.7–74.0)	Fresh gale	Twigs break from trees; difficult to walk
9	47–54 (75.6–86.9)	Strong gale	Chimney pots and slates blown from roofs; slight damage to structures
10	55–63 (88.5–101.4)	Whole gale	Considerable damage to buildings; trees uprooted
11	64–75 (103.0–120.7)	Storm	Widespread damage
12	more than 75 (120.7)	Hurricane	Devastation

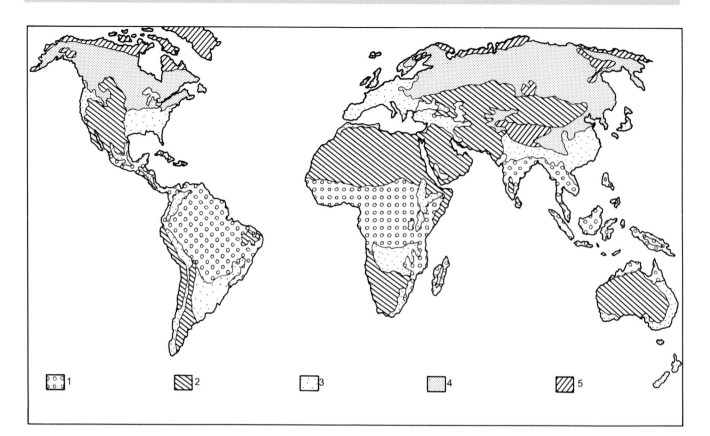

 has this key below:

1 ◦◦◦   2 ▨   3 ⬚   4 ▦   5 ▨

The main climate groups of the world.

Key
1 Tropical rain climates
equatorial: rain all year (rain forest)
tropical: rainy and dry seasons (savanna) grassland

2 Dry climates
arid: desert
semi-arid: sparse rain (steppe grassland)

3 Warm temperate rain climates
warm-temperate, dry in summer
warm-temperate, dry in winter
warm-temperate, rain all year

4 Cool temperate rain climates
cool-temperate, severe winters, rain all year
cool-temperate, severe winters, winters dry

5 Polar climates
polar: cold all year

August — although it is markedly colder at Thule, in the north. Precipitation is moderate — about 30.5 inches (775 mm) in an average year.

Iceland lies in the same latitude as southern Greenland, and very close to the Arctic Circle, but it is distinctly warmer. At the capital, Reykjavik, the average temperature is at freezing in January and February, which are the coldest months but, in July and August, it averages 52 °F (11 °C). Precipitation over the year is much the same as that of Greenland. The difference occurs because during the summer, Iceland is within a mP air mass but, in winter, it lies close to the front between mP air to the west, cA air to the north and, more significantly, a region of mixed air to the south. This area, which also covers Britain, is where tropical, polar and arctic air meet. Their interaction produces many local areas of low and high pressure which are constantly moving and, consequently, weather that is often unsettled but, on the whole, mild.

Britain, and in summer much of western Europe, lie in this fairly small area of mixed air. Should the front between the mixed air and the air to the south move just a little northwards in winter, Britain — or the southern part of it — will lie within another

area of mixed air, where mT to the west, cT to the south, and cP to the east meet. In summer, a northward movement of the front will place Britain within the Eurasian cT air mass. Iceland is too far north ever to lie beneath tropical air but it, too, has a changeable climate.

In the world as a whole, these conditions are unusual. Most people live in places where the climate is dominated by one, or at most two, types of air mass. Provided they live near its centre, weather forecasting presents them with few difficulties. In Central Asia, for example, or in the central part of Australia, the weather is fairly predictable from one month to the next, although exceptional weather can occur in any season. The meteorology grows more complicated near the boundaries between air-mass source regions, where air of one type is moving into a region dominated by a different type and its characteristics are being altered.

Local topography can also produce weather that is not typical of the large region in which it occurs. Mountain ranges, in particular, produce local climatic variations. By making it rise, they can 'strip' arriving air of its moisture, for example, so much more precipitation falls on one side of the range than in the 'rain shadow' on the other. Even then, the conditions are usually predictable provided the movements of air are reliable and always involve air masses of the same types. A coastal belt or mountainside that receives heavy rain is likely to do so year after year. Air masses are good guides.

Difference between climate and weather

Wladimir Peter Köppen (1846-1940) was born in St Petersburg (Leningrad) of German parents. He grew up to become one of the leading meteorologists of his generation, working first in Russia and later in Germany. His most important contribution to the scientific study of weather, and the work for which he is best known, was his classification of types of climates, which he produced and continually developed and refined between 1900 and 1936.

Our weather is brought to us by air masses. We can describe the characteristics of each type of air mass, and this will give us at least an approximate idea of what weather to expect, but it is also possible to turn the whole concept around. Instead of looking at climate from 'the top down', we can look at it from 'the bottom up' and consider more precisely the conditions associated with it. Such information is of great importance to agricultural scientists, who need to know what crops can be grown in a particular region, and to botanists, zoologists and ecologists, who need to know what species of plants and animals are likely to occur.

This is the approach Köppen adopted. He defined six types of climate mainly on the basis of temperature. A tropical rainy climate is one where the temperature in the coldest month is higher than 64.4 °F (18 °C). His second category comprised dry climates. The third are the warm, temperate, rainy climates in which temperatures in the coldest month are between 26.6 °F and 64.4 °F (-3-18 °C) and in the warmest month more than 50 °F (10 °C). A cold boreal forest climate — in which conifer forests are the dominant type of vegetation — has temperatures of less than 26.6 °F (-3 °C) in the coldest month and of more than 50 °F (10 °C) in the warmest. In a tundra climate the temperature in the warmest month is between 32 °F and 50 °F (0-10 °C). A perpetual frost climate has a warmest month in which the temperature does not rise above freezing.

Köppen based his classification on the extremes of temperature different types of vegetation can tolerate. Certain tropical plants will die if the temperature falls below 64.4 °F (18 °C), for example, and trees will not grow if the temperature never rises above 50 °F (10 °C).

The classification is crude and there are many exceptions to the vegetation limits on which it is based, but it is still widely used. Other scientists sought more reliable classification systems, however. Some related climates directly to the air masses which deliver them. The German climatologist, H. Flöhn, for example, based a system on the global wind belts and the precipitation associated with them and, in 1969, A.N. Strahler produced what may be the simplest system of all. He divided all the climates of the world into three broad types. Equatorial and tropical air masses produce low-latitude climates, middle-latitude climates are produced by tropical and polar air masses and high-latitude climates are produced by polar and arctic air masses. Strahler then subdivided these groups into 14 climatic regions and added highland climates, so there were 15 types in all.

Other scientists approached the problem differently and what is probably the classification most widely used by modern climatologists was devised by an American climatologist, C. Warren Thornthwaite (1899-1963), and first published in 1931 and revised in 1948. Thornthwaite was concerned with the water requirements of farm crops and his system relates precipitation (P) to evaporation (E) by measuring the amount of precipitation in each month and dividing it by the amount of evaporation. Added together, the resulting figures for each month of the year produce a 'P/E index' that defines a climate and, to some extent, the vegetation it supports.

The original Thornthwaite system defines 'humidity provinces'. The 1948 modification relates the water needed by particular types of vegetation to the precipitation available to them and also introduces temperature, by adding together the average temperature for each month to produce 'accumulated temperatures' ranging from 0, which indicates a frost climate, to 127 for a tropical climate.

No matter how you classify types of climate, they will correspond closely to the vegetation typical of the regions in which they occur. For general purposes, geographers usually take elements from several classification systems — but especially from Köppen — to divide the climates of the world into 10 types, subdivided into five groups. The first group comprises tropical rain climates, subdivided into those that have rain throughout the year and those where distinct dry and rainy seasons occur. The second group includes the dry climates, subdivided into deserts, effectively without rain, and semi-arid grasslands — the steppes of central Asia, for example, and the African savanna. The warm temperate rain climates are a little more complicated. The group includes those which have a dry, hot summer, those which have a dry winter and short, cool summer and those in which rain occurs throughout the year and where the climate may be warm or cool. The dry, hot summer type occurs around the Mediterranean, for example, and the type which has rain throughout the year and hot summers in the eastern United States. Cool temperate rain climates include those which have rain throughout the year and a severe winter and those — found mainly in north-eastern Asia — which have a cold, dry winter. Polar climates, forming the final category, have no warm season.

Weather forecasting

Classification of this kind makes it possible to draw a map of the world that shows the general climate over large areas. It is simple but it is no more than a generalization, a broad outline based on what you may expect in a typical year. It says very little about the actual weather you may expect tomorrow or next week. The study and classification of the climates of the world is the subject of its own scientific discipline, 'climatology'. The related study of actual weather, and its forecasting, is called 'meteorology'.

As the last ice age approached its end and the Sun began to shine more warmly, making the glaciers and ice sheets retreat, people in several parts of the world began to change their way of life. Instead of hunting wild animals and gathering wild plants, they set about domesticating livestock and growing crops. Human well-being had always been linked to the weather, but the introduction of farming made it important to predict when conditions would be suitable for planting and harvesting. When sailors began to venture further, in search of fish or trade, their lives depended on an ability to predict storms and fair winds.

People have always gazed at the sky looking for signs, and it was in classical times that the first attempts were made to introduce some kind of scientific method into studies of the weather, and its prediction. Aristotle devoted a work called *Meteorologica* to the subject. The Greek word *meteoros* means 'lofty'. It is the word Aristotle used and our word 'meteorology', literally the study of the sky, is derived from it.

Progress was slow, because without a systematic record of measurements it is impossible to track the movement of weather conditions and, unless their past movements are known, there is no way of telling where they are heading. For centuries people had to make do with their observations of nature or with rough guides, 'rules of thumb' that were often reduced to catch-phrases or snatches of doggerel to make them easier to remember. Some of these were reliable, others were not — but it is human nature to remember their successes, and forget their failures, so they persisted.

A 'red sky', for example, is caused by dust in the air, which scatters blue light but not red. It indicates dry air and, therefore, fine weather. See it in the evening ('at night' to rhyme with 'delight') and it is to the west and approaching, though it may pass during the night. See it in the morning ('warning') and it is to the east, which means the dry air has passed — weather moves generally from west to east — and wet weather may be following it — but, then again, it may not. When warm air is forced to rise over cold air, so its moisture condenses, clouds are likely to form up the slope of the front. The approach of high-altitude clouds, therefore, often heralds the arrival of a front and rain. Other weather lore, however, is based on less secure foundations. The size of the crop of berries in late summer, for example, or the presence or absence of particular species of animals, may well be a consequence of the weather that has passed, but tells us nothing about the weather to come. Plants and animals react to the conditions they experience, but they are quite unable to foretell the future. Predictions based on such phenomena are really based on our sense of natural justice. If the weather has been bad, we feel we are 'owed' something better, and if it has been good (producing many berries,

Meteosat photographs of the world show clearly the distribution of cloud cover. Note the thick cloud over Equatorial West Africa and the clear skies over the Sahara and Arabia.

perhaps), we feel we will be 'made to pay for it' during the coming winter.

Some people make long-term predictions, for a season or even a year ahead, which they base on past records. They analyse the weather conditions over the preceding weeks or months, search past records for as many similar sequences of weather at that time of year as possible, read from the records what happened next and try to detect a pattern which they can use as a prediction of what seems most probable. The method is very hit-and-miss, but may have a partly sound basis and professional meteorologists use it — but only as one of the many methods available to them. The link between the occurrence of an ENSO event (see page 92) and widespread climatic events during the subsequent year is now well established. This allows very general predictions to be made, at least in principle, and the very early stages of typically 'El Niño' or 'anti-El Niño' weather might be

discernible as a characteristic series of events — a 'signal'.

Meteorology

The scientific development of meteorology began in the seventeenth and eighteenth centuries, with the invention of accurate instruments. Galileo Galilei (1564-1642) invented the thermometer in 1607, and Evangelista Torricelli (1608-47) invented the barometer in 1643. Theorists, who sought to understand the behaviour of fluids, including air and water vapour, followed and at one time or another most of the greatest scientists of the eighteenth and early nineteenth centuries turned their attention to the weather. It was during the nineteenth century that observing stations were established where standard measurements could be made regularly and recorded. During our own century, the network of surface observing stations has been increased and, as the extent of the climatic

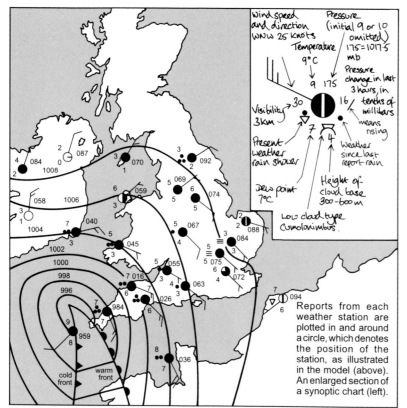

Synoptic chart showing reports from weather stations.

Orbiting satellites transmit instrument measurements and also photographs which show the type and distribution of clouds. Two types of satellite observation are used. The first comes from 'polar orbits'. As the name suggests, a polar orbit carries the satellite from pole to pole, one complete orbit taking it 100 minutes. As it moves, the Earth is rotating beneath it — in 100 minutes the Earth turns through 25° of longitude — and so each orbit sweeps a different path. The other observations are from satellites in 'geostationary' orbits. If a satellite is placed in an orbit over the Equator at a height of 22,000 miles (35,398 km), its speed will be the same as that of a point joining the satellite to the centre of the Earth through a particular surface location and rotating with the Earth. In other words, the satellite will remain permanently above the same place on the surface. Geostationary satellites transmit a complete scan of the whole disc visible to them at a rate of one every half hour.

The surface stations collect standard measurements four times a day and send them to one of a network of receiving stations four times a day. The observation times are standardized throughout the world, as midnight, 0600, noon and 1800 hours Greenwich Mean Time. They record the wind speed and direction, temperature, visibility, present weather conditions, the weather since the last report, the dewpoint temperature, the type of cloud and height of its base, the total amount of cloud cover, the pressure, the amount by which the pressure has changed since the last report and whether it has risen or fallen.

At the receiving station this information is plotted on a map. The location of the observing station is marked and the information from it is represented by a set of standard symbols and numbers arranged formally around the circle representing the station itself. The result, when the incoming measurements from all the stations have been plotted, is a map that summarizes the weather conditions over the entire area being covered. A summary is a 'synopsis', and a map summarizing the weather is called a 'synoptic chart'. The first synoptic chart to be compiled in Britain was displayed at the Great Exhibition of 1851. Synoptic charts are also compiled from observations made in the upper atmosphere, each chart corresponding to a particular altitude.

influence of events in the upper troposphere and stratosphere came to be recognized, observations of the upper atmosphere have been introduced. At first, these were obtained from instruments attached to kites and balloons but, more recently, high-altitude aircraft, rockets and orbiting satellites have been used.

The purpose of the observations is to gather as much precise information as possible from the widest possible area. This includes sea areas as well as land — and the weather at sea is most important in western Europe, where weather systems arrive from the Atlantic. A small number of weather ships are moored in particular locations — the British Meteorological Office has about 300 of them — and commercial vessels and aircraft report the conditions they encounter but, obviously, these relate only to the regular shipping lanes and airways. Radiosondes and rawinsondes provide information about conditions at heights of up to about 40,000 feet (12 km). A radiosonde is a balloon that is equipped with instruments and a radio transmitter. As it rises, at about 16 feet per second (5 m/s), it broadcasts the instrument readings to the ground. A rawinsonde is a radiosonde that is tracked by radio or radar to provide details of wind speed and direction.

The synoptic charts are converted into the more familiar 'weather maps' by drawing lines to join together stations that report similar atmospheric pressure. The lines are 'isobars', and the additional information from the observing stations allows the meteorologist to plot the location and type (cold, warm, or occluding — the stage in the development of a depression where warm air is being lifted over cold air) of fronts.

All of this can then be checked against the satellite observations and photographs. These provide, for example, a broad picture of the location and type of cloud to augment the surface reports. The surface stations themselves report the type and amount of precipitation falling from the clouds and such features as thunderstorms.

Having completed all this study, and armed with the most recent synoptic charts, the meteorologist is in a position to take the big step of predicting how the picture will change over the forecast period. Time is not on the side of the forecaster — a forecast is of little use if it is overtaken by events during the time taken to prepare it.

At one time, this was a matter of skill, judgement and experience, because what matters, and is difficult to predict, is the direction in which the overall system is moving, its speed and the extent to which the characteristics of the air mass are changing as it moves. Skill and judgement are still in demand, but the modern meteorologist employs yet another tool to assist — the computer.

As information arrives, and while it is being plotted on the synoptic charts, it is translated into numerical values and fed into a computer program. The program consists of a vast number of mathematical equations and formulae that describe the behaviour of gases under particular conditions of temperature and pressure. It contains values that correspond to the last set of observations fed into it, so each new set of numbers represents a modification of those it already has. By applying them to its equations and formulae it is able to calculate the amount and rate of change since its last run and then, by assuming the rate of change remains constant, to extrapolate into the future — thus producing a forecast. When you watch a television weather forecast, and the forecaster describes the clouds, fronts and shadows crossing the map as 'this is what the computer predicts', this is the process that underlies the prediction.

In addition to the general forecasts, supplied to broadcasters and newspapers, more specialized forecasts are provided for

A modern meteorological station, on board the British weather ship *Cumulus*.

those who need them and, while a mistake in a general forecast may cause inconvenience, a mistake in a specialist forecast could lead to tragedy. Farmers, sailors, mountaineers and hill walkers all require particular kinds of information, usually concerning a very local area. Aircrews — civil and military — need to know about the visibility at ground level and height of the cloud base, for taking off and landing, and about conditions, especially wind speed and direction, at various heights.

People make jokes about the supposed inaccuracies of general forecasts but, in fact, they have a reasonably good record and meteorologists couch them in language that is sometimes deliberately vague, because there will always be local exceptions in a description that is generally accurate — it is possible to forecast showers, for example, but not precisely where each one will occur. Yet, despite the power and sophistication of its technology, weather forecasting can peer for only a short time into the future. The behaviour and movement of regions within air masses can be affected by quite small events and the significance of small changes can grow quickly. A deflection of a few degrees in the track of a hurricane, to take an extreme example, can determine whether or not it strikes a populated area.

Climatology

Meteorologists are concerned with the weather — the particular phenomena that are experienced and the mechanisms which produce them. Climatologists are concerned with the general types of weather that occur over large regions of the world, and are typical of conditions within those regions over long periods of time. The two branches of the atmospheric sciences overlap, obviously, since a scientist working in one must have a good understanding of the other, but they operate on different scales. A climatologist is interested in the behaviour of whole air masses, affecting entire continents and oceans, and in time scales covering thousands, or even millions, of years. If there is such a thing as a climatological equivalent of a weather forecast, it is a climatological prediction of the general weather conditions that may prevail decades, centuries or even millennia from now.

The climatologist approaches the task from two directions, one mathematical, the other historical. Mathematics are involved — and are formidable — because the behaviour of gases under different conditions of temperature and pressure are described by the 'gas laws', which are expressed as equations, some of them involving fairly simple algebra and others, which are needed for fluids that are in constant motion, require calculus. The historical approach is designed to reconstruct climates of the past — often of the remote past long before humans recorded their experiences or before humans existed at all.

The subject matter of climatology is called the 'general circulation' (or sometimes 'global circulation'). This is the technical term — the 'shorthand' — that describes the overall composition and movement of the atmosphere. The Hadley cell circulation, the overall transfer of energy from the Equator to the poles, the jet streams, and the formation, movement and modification of air masses all form part of the general circulation. Year by year, climatologists collect information from meteorological observations made all over the world and assemble it to build a comprehensive picture of the way the global atmosphere develops and delivers its climatic conditions. Then, using their knowledge of the gas laws, they attempt to discover how the atmosphere may behave in the future and how it may react to perturbations.

A 'picture' of the general circulation is known technically as a 'model'. It is an attempt to reduce a vast and extremely complex aspect of the real world to a description that is simple enough to be comprehensible, but not so over-simplified as to be misleading. A model is intended for use. It has an obvious value in teaching and all scientific education relies extensively on simplified descriptions — models — of the natural world. It is also of great research value because, if it exists in a suitable form, a model can be manipulated. Climatologists use experimental techniques, as do all scientists, but they are able to experiment only on their models — it would be impossible, and most unwise, to manipulate the atmosphere itself to observe its response.

A model may be verbal — a spoken or written description, such as a teacher or textbook author might use — or mathematical. Research models of the general circulation, called 'general circulation models' or

'GCMs', are mathematical. They exist as equations, to which values can be attached, and they are constructed and operated with the help of computers.

In principle, the construction of a GCM is not especially difficult for anyone skilled in computer programming. The modeller begins by designing a set of grid lines that cover the whole world. These are rather like lines of latitude and longitude but, because the atmosphere is three-dimensional, the grid must produce cubes rather than squares — sets of 'latitude and longitude' grid lines are stacked one above another. All the lines in the grid are labelled so that any position within the grid can be identified by a set of coordinates.

The next step is to add all the gas-law equations in the form of an algorithm — a sequence of instructions that directs the computer to perform calculations in a specified order. The grid and equations are related to one another by an overall program. This instructs the computer to store and use the information supplied to it to construct a mathematical description of the condition of the atmosphere at a particular time and then, as that information is replaced by new information, to calculate the consequences of the changes and construct an appropriate description of the new atmospheric conditions. The model

may then be allowed to run by itself, using the results from one sequence of its calculations to supply the starting conditions for the next, and so demonstrating the evolution of the global climate over a period of time.

The initial information is derived from actual observations and measurements — from meteorological stations, rawinsondes or satellites, for example. When the scientists wish to discover the consequences of perturbations, they supply the information themselves. They might wish to discover what would happen if the Sun shone more or less brightly, for example, or if the chemical composition of the atmosphere were to change.

The difficulty of climatic modelling results from the size of the system the models describe and, therefore, the large number of calculations that are involved. In each run of the model, the equations are solved for every point at which grid lines intersect. The grid lines must be as close together as possible, because many climatic features are relatively small and widely space intersections may miss them. A forest in what is otherwise farmed land, or a large lake, for example, may affect humidity locally, and clouds, which influence the humidity, lapse rate and surface temperature, are often no more than a mile or two

The global carbon cycle General circulation models must take account of all the factors which can affect climate, one of which is the movement of carbon dioxide into and from the air. This diagram greatly simplifies the situation in the real world, but illustrates the complexity of the problem. Amounts of the carbon that is moving between different parts of the cycle are shown in round brackets (); carbon that is held in 'store' in square brackets []. The units used are multiples of a gram (1g = 0.035 oz).

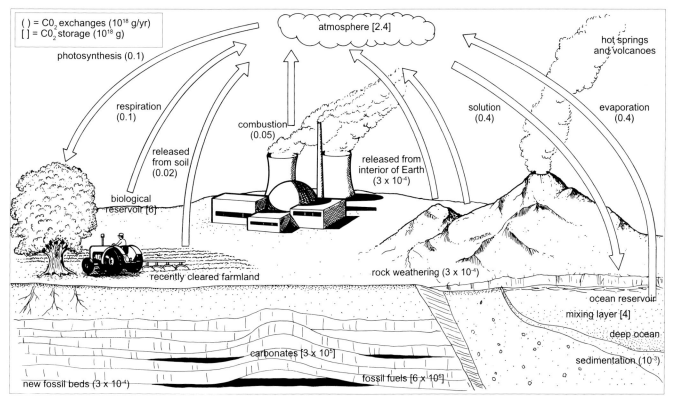

( ) = CO_2 exchanges (10^{18} g/yr)
[ ] = CO_2 storage (10^{18} g)

atmosphere [2.4]

hot springs and volcanoes

photosynthesis (0.1)

respiration (0.1)

combustion (0.05)

released from soil (0.02)

solution (0.4)

evaporation (0.4)

released from interior of Earth (3 x 10^{-4})

biological reservoir [6]

recently cleared farmland

rock weathering (3 x 10^{-4})

ocean reservoir

mixing layer [4]

deep ocean

sedimentation (10^{-3})

carbonates [3 x 10^5]

fossil fuels [6 x 10^6]

new fossil beds (3 x 10^{-4})

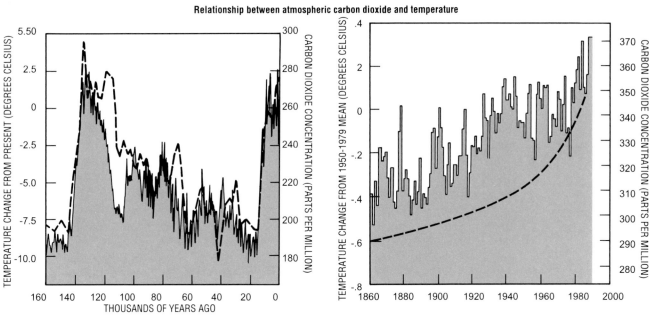

Relationship between atmospheric carbon dioxide and temperature

across. The finer the grid, however, the larger the number of its intersections and the more calculations that must be performed — and each calculation takes a certain time. This means that the greater the computing power available to it, the more reliable a GCM will be. In practice, GCMs are constructed only on the very largest supercomputers. Even then, they are relatively crude and, in their finest grids, the intersections are about 62 miles (100 km) apart. The large computers that are now being developed will offer great improvements in computing power and speed that will allow GCMs to use finer grids — thereby increasing their sensitivity and reliability.

GCMs are also limited by the fact that certain very important climatic influences are not well understood. Scientists have a general understanding of the ways clouds form, for example, but many details of the processes involved have not yet been resolved. The oceans are known to affect the air in contact with them, but there is much uncertainty about precisely how that influence is exerted. The resolution of such questions must await the results of further research. For the time being, modellers have to make do with general assumptions about processes that are not understood in detail.

The reconstruction of climates in the remote past relies on a wide range of techniques 'borrowed' from almost as wide a range of scientific disciplines. The identification of pollen extracted from soil samples,

for example, reveals the plants which grew in a particular place and the radiometric dating of organic fragments in the soil reveals when. Once the type of vegetation is known, it is possible to describe the climate in which it grew.

Ice cores, obtained by drilling into the Greenland and Antarctic ice sheets, also provide valuable climatic information. The ice sheets form as, year after year and century after century, snow falls on top of earlier snow. Little by little, the weight of snow compresses the lower layers until the snow becomes ice, but the snow which fell each year forms a thin but distinct layer. By counting, the layers it is possible to work out approximately when particular sections of a core were formed. Fresh snow lies loosely on the surface, with tiny air spaces between the crystals. As the snow is compressed, some of this air is trapped, as bubbles in the ice.

Scientists remove samples of this air, being very careful not to contaminate it with modern air, and analyse its composition. The analysis will reveal, for example, how much carbon dioxide or methane the air contained at a particular time in the past. The ice itself records the climate at the time the snow fell. The thickness of the layer indicates the amount of precipitation and the relative proportions of two isotopes of oxygen, O-16 and O-18, indicate the temperature. The isotopes are identical chemically, but the higher the proportion of O-18, the lower the temperature of the water.

(Left) The ratios of two isotopes of oxygen taken from ice cores of known age have been used to calculate the temperature over the last 160,000 years. When plotted against the atmospheric concentration of carbon dioxide, which can also be measured, the two sets of values follow one another very closely. (Right) Similar values from 1860 are less closely correlated and it is uncertain whether the slight warming is linked to the accumulation of carbon dioxide in the air. See also pages 160-161.

(Opposite) The Climatic Research Unit at the University of East Anglia, Norwich, England, where computer models are made of the general circulation of the atmosphere.

CLIMATES OF THE PAST

If you look at the end of a log that has been sawn from the trunk or bough of a tree, you will see that it bears a pattern of concentric bands of dark and light colour. These are tree rings. In spring, the tree begins to grow new wood on the outside of its trunk, just below the bark. The new growth consists of large cells with thin walls, interspersed with the vessels through which water moves up and down the tree. From about the end of summer, although growth continues, the new cells are smaller, have thicker walls and, in some species, there are fewer vessels. In a cross-section through the trunk, the large cells appear pale and the smaller ones dark. The pair of rings represents the growth during one year and the dark rings separate each spring growth from the next. You can date the tree by counting the dark rings, one ring for each year.

A skilled observer can tell more. Thick rings indicate abundant growth and thin rings poor growth. The rate at which a tree grows depends mainly on the amount of rain and sunshine it receives. If you know the environmental requirements of the tree concerned, therefore, these differences can be interpreted to reveal something about the weather, year by year. The dating of trees by counting their rings is called 'dendrochronology', and placing a climatic interpretation on them is 'dendroclimatology' (the Greek *dendron* means 'tree').

The dendrochronological record is not confined to the recent past. The logs we see around us — waiting to be placed on the fire, perhaps — were alive recently and most trees live for no more than a century or two. Not all trees are so short-lived, however. Bristlecone pines (*Pinus aristata*), which grow in a very dry part of California, live for thousands of years and their rings can be examined by taking very slim cores which do no serious harm to the tree. Some bristlecone pines have lived for 4600 years — and are alive today — and dead trees, which do not decompose in that environment, provide a continuous record going back more than 8000 years.

Even trees that have been dead for a very long time can be used to reconstruct the climatic conditions under which they grew. The age of wood can be determined by radiocarbon dating, and so their rings provide a precise historical record of the period during which they lived. Ancient logs are sometimes preserved in peat bogs, but a cross-section from wood that has been used for building or to make household articles, which can be dated by historical records, will serve just as well.

The importance of historical records

The historical records, augmented by archaeological discoveries, also chronicle the history of our climate. In the seventh century, busy roads crossed the Alps, through passes that are now filled with ice, and the Viking settlement of Greenland, in the tenth century, took place when the climate there was warmer than it is now. The Greenland settlements were abandoned early in the fifteenth century, when conditions deteriorated, at the beginning of the 'Little Ice Age'.

During this cold episode, lasting from the middle fifteenth to the middle nineteenth centuries and affecting most of the northern hemisphere, glaciers advanced and, in the Alps, some mountain communities were forced to move to lower levels. Contemporary accounts tell us what the weather was

A section through the trunk of this black locust tree (*Robinia pseudoacacia*) reveals the annual growth rings.

Needles and cones of a bristlecone pine (*Pinus aristata*).

like. Novels written as late as the early nineteenth century describe British winters in which it was commonplace for even large rivers to freeze over and snowfalls were heavy. The tree rings and glaciers confirm what we might otherwise be tempted to regard as 'poetic licence'. Since about 1850, glaciers have been generally retreating, with a few pauses and brief advances, as the Little Ice Age is left behind.

Technically, the remains or traces of plants and animals that lived more than 10,000 years ago are 'fossils'. Fossils are often mineralized — the organic remains have disappeared and the space they occupied filled by a mineral — but this is not always the case. Ancient bones are sometimes found and even whole animals. From time to time, mammoths are discovered in Siberia, preserved by the very cold conditions.

In the case of species that still exist, we know the climatic conditions they can tolerate and, for extinct species, it is often possible to infer them. Teeth, which are often preserved, give a good indication of the diet of the animal from which they came and, therefore, of the way it lived. Mammoths, which became extinct about 12,000 years ago, were very large and well covered with hair. It is reasonable to conclude that they lived in cool regions and on open grassland or tundra — conditions similar to those over much of northern Asia today.

At one time, mammoths were widespread and their remains are often found not far from those of horses which are also grassland animals of temperate or cool climates. Sometimes, though, there are surprises and species are found that could not possibly tolerate the conditions which obtain today.

Such finds usually arouse little public interest but, occasionally, the location of the find makes it something of a sensation. A find of this kind was made in the very heart of London (and there has been a

In the past, climates have been both warmer and cooler than they are today. During part of the Ipswichian Interglacial, for example, what is now central London had a tropical climate and the fossil remains of hippopotamuses and elephants have been found there.

similar find near Cambridge). Over the years, the course of the River Thames has changed and buildings now stand on what was once the sandy river bed. Building excavations now and then expose the deposits — and their contents. This is how remains were discovered, beneath Trafalgar Square, of animals and plants that can tolerate only conditions much warmer than those Londoners enjoy now. There were fossil remains of hippopotamuses, rhinoceroses (not the woolly species of cold climates, which is now extinct) and of a species of elephant, together with remains of invertebrate animals that now live far to the south of Britain and of what today are southern European and North African plants — such

as water caltrop (*Trapa natans*) and Montpellier maple (*Acer monspessulanum*). These species lived, where central London stands now, about 100,000 years ago, when summer temperatures must have been up to 5 °F (2.8 °C) warmer than they are now.

This sounds a very small difference but Bordeaux is the nearest city with a July average temperature 5 °F (2.8 °C) warmer than London. Changes in average temperatures even smaller than this can have, what may seem to be, quite disproportionate climatic consequences. Around 1700, for example, when the Little Ice Age was at its coldest, the average temperature in Britain was only about 2 °F (1 °C) colder than it is now.

The hippopotamuses and rhinoceroses inhabited Britain during one of the warm intervals — interglacials — which interrupted the series of ice ages that occurred over about the last 2 million years (see page 130), and about which we know a great deal. Not long before the start of the ice ages, about 3 million years ago, certain fossils suggest that the world must have been much warmer than it has been at any time since. The fossilized leaves of southern beech trees (*Nothofagus* species) have been found in Antarctica, lying in a large heap, just as they fell to the floor of the beech wood in which they had been growing. The wood was in the Transantarctic Mountain Range, which divides the continent in two, and only 250 miles (400 km) from the South Pole — on what is now the Beardmore Glacier. The known tolerances of southern beeches imply that, at the time they were alive, summer temperatures rose to about 41 °F (5 °C) –27 °F (15 °C) warmer than today — and in winter they fell to about -4 °F (-20 °C).

Fossils

Move back much further in time and the fossils tell another story. Ordinarily, when plants die, their tissues decompose but, occasionally, decomposition may be arrested. This is most likely to happen when plant material falls into shallow, fairly stagnant and rather acid water containing very little dissolved oxygen and is buried quickly beneath mud. After a time, the vegetable matter will turn into peat. If it is buried deeply, however, and some time later subjected to high temperatures and pressures, the peat will be converted into one or other of the many types of coal.

Coal is by far the most abundant of the world's 'fossil' fuels. They are called 'fossil' fuels because originally the word 'fossil' meant 'dug from the ground' and had no implication of great age or of organic origin. Coal was simply a kind of rock that could be dug from the ground and that would burn. It occurs in seams, separated by other kinds of sedimentary rocks. This suggests that the environment in which it formed was not constant. A period of conditions suitable for coal formation would be followed by a period of different conditions.

Apart from Antarctica, which has not been explored to any extent, coal is found in every continent and usually in very large

The characteristic winding gear is the only evidence above ground of the deep coal mine below, here in the English Midlands, that supplies fuel to the electricity generating station behind.

amounts. The United States has estimated reserves of more than 3000 billion (thousand million) tons, China may have in the region of 900 billion tons, the USSR more than 400 billion tons and Britain has more than 40 billion tons.

The plants that died and eventually were converted into coal grew at various times in the past. Some lived more than 400 million years ago, but most were growing about 300 million years ago. They grew, all of them, as swamp forests and such forests occur only in the kind of warm, humid climates we associate today with Equatorial regions. Clearly, the forests extended over vast areas — across much of what are now North America, Europe and Asia, for example — and this suggests that the Earth has experienced several periods when the climate over these large areas was a great deal warmer than it is now.

Devon, in the south of England, is famous for its red sandstone cliffs and the fertile red soil that has formed from them. The region is also famous geologically and has given its name to a geological period, the Devonian. About 400 million years ago — between the earlier and later episodes of coal formation — Devon was a dry desert, much like the modern Sahara, bordering a long-vanished sea. Today, Devon has a mild, rather wet climate. Its lush pastures feed dairy cattle and its coastline and red cliffs attract summer tourists.

There can be no doubt that the climates of Earth have changed over the years, and there is every reason to suppose they will continue to do so. Some changes have been

(Opposite, top) An alpine tundra landscape, in Colorado. (Opposite, bottom) Pine forest, in Germany, that is similar to the high-latitude boreal forest.

gradual, but others have been sudden — an ice age can begin or end within a century or less and, on a smaller scale but one that had a considerable effect on the people who lived during it, the Little Ice Age began and ended abruptly and without warning.

The changes rarely involve more than a small increase or decrease in average temperature. The differences between our present climates and those of the remote past, however, should be interpreted with caution. We have theories to explain the onset or termination of an ice age, but it is unwise to apply them to the conditions that produced the coal-forming swamps or the desert that once was Devon. It is more likely that the swamps and deserts were not always where they are today.

Swamp forests once covered large areas of all the continents and, therefore, at one time the climates of those continents must have been warm and wet. This could imply that at those times the entire global climate was much warmer than it is now, or that the continents then lay much closer to the Equator or, perhaps, some combination of the two.

Continental drift

If you examine a map of the world, it may occur to you that some of the continents look as though they should fit together, like pieces of a jigsaw. Africa, in particular, seems to fit neatly against Central and South America. Many scientists have made the same observation, but it was the German meteorologist and geologist Alfred Wegener (1880-1930) who drew together a great variety of detailed observations and, in about 1910, proposed a theory of 'continental drift'. Wegener suggested that throughout most of the history of the Earth there was but one land mass. This great continent, which came to be called Pangaea, straddled the Equator, extended to the poles, and occupied about one-third of the surface of the planet. It was shaped rather like the letter U, lying on its side, the northern and southern parts surrounded by a great ocean, Panthalassa, and separated by a smaller sea, called Tethys. Then, about 200 million years ago, the two parts of Pangaea began to separate.

A further 60 million years or so passed, and then the two 'supercontinents' — Laurasia in the north and Gondwana in the south — began to fragment. The names of the supercontinents were coined in the 1930s by a South African geologist, James du Toit (1878-1948). Laurasia separated

During the Carboniferous Period, about 360 million to 286 million years ago, swampy lowland areas supported forests of plants that are now extinct, including *Lepidodendron* (foreground, fallen), which sometimes grew to a height of 100 feet (30 m) before the first branch, and *Calamites*, a relative of the modern horsetail, which grew to a similar height. The remains of these forests formed many of the coal measures we now mine.

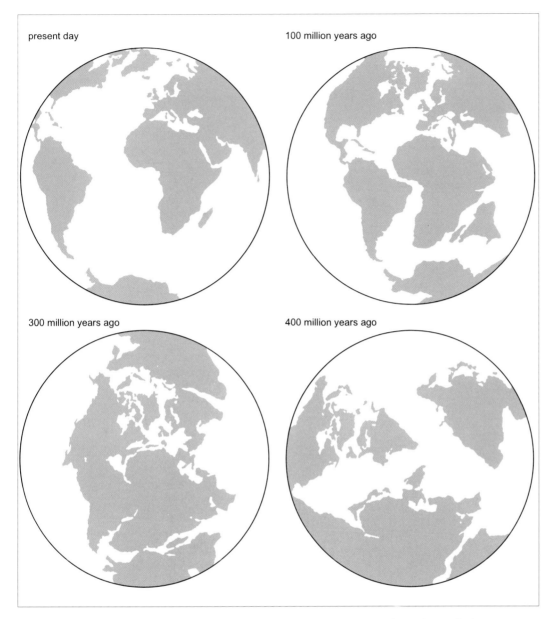

present day

100 million years ago

300 million years ago

400 million years ago

At the time of the great coal-forming swamps, about 400 million years ago, most of the land on Earth formed a single continent, Pangaea (the land in the north-eastern part of the map is part of Pangaea, which extended right around the world). Pangaea was surrounded by a great ocean, Panthalassa, and the smaller Tethys Sea separated its northern and southern regions. About 300 million years ago, Pangaea was just beginning to separate to form two continents, Laurasia in the north and Gondwana in the south. By 100 million years ago Laurasia and Gondwana had become still more divided and the present-day continents can be identified clearly on the map. The Tethys Sea remained between Eurasia and Africa. Today, Africa has joined Asia, and the Mediterranean is all that remains of the Tethys Sea, while a new rift has appeared, forming the Red Sea between Africa and the Arabian peninsula.

eventually into what are now North America, Greenland, Europe and Asia. Gondwana became South America, Africa, India, Australia and Antarctica. The movement continues to this day. The North Atlantic, for example, did not begin to open until about 80 million years ago, and it is still growing wider.

The theory of continental drift explained the location of coal deposits — they were formed near the Equator on land that has since drifted into higher latitudes — and the presence of certain plant and animal species on continents that are now widely separated. Biologists were impressed, but geologists could think of no way in which large sections of the Earth's crust could move independently of one another. It was many years before the theory of plate tectonics

was devised to explain the underlying mechanism.

Continental movements affect climates profoundly, because they alter the distribution of air masses and, therefore, the general circulation of the atmosphere. The southern ice cap exists because Antarctica, the source of a large continental air mass, has drifted to the South Pole. The Himalayas, still rising because India is moving northwards, colliding with Asia, prevent moist air from reaching much of India.

The causes of ice ages

Imagine what would happen if, one year, the average summer temperature were just a degree or two lower everywhere north of, say, New York. Some distance further to the north, in Canada, some of the snow that

(Opposite, top) Open, deciduous woodland has areas where daylight reaches the ground and allows grasses and herbs to grow. This type of park-woodland, here in the Great Smoky Mountains, Tennessee, provides shelter and pasture for grazing animals. (Opposite, bottom) Mountain avens (*Dryas octopetala*), a plant of alpine regions and high latitudes. Its presence as a fossil, or sometimes as a living plant, indicates that places which are now too warm for it had a cooler climate in the past.

had fallen in the preceding winter would fail to melt. Snow would cover ground that had been free from snow in previous years and, because it is white, it would reflect sunlight. This would prevent the ground from absorbing warmth and, at the same time, the snow would chill air that passed over it. The following winter, more snow would fall on ground that had been prevented from warming during the summer. The following summer, the air would remain cold, because of the snow, and the area covered by summer snow would increase. The entire process would develop as a kind of vicious spiral, feeding on itself. This is called 'positive feedback' and it affects air and, therefore, climate strongly.

It can happen only if there are large land areas in high latitudes. Snow that falls into the sea melts at once and the sea does not freeze readily. The freezing point for salt water is much lower than for fresh water, and winds and currents keep the sea surface in constant motion, which inhibits the formation of ice. The sea freezes only where it is exposed to continental polar air — air that has been severely chilled over land. There is no land at the North Pole, but the ice cap is surrounded by continents on all sides.

The feedback mechanism also works the other way, just as strongly. Suppose there is a series of unusually mild winters, with no change in summer temperatures. Snow would cover a smaller area in winter and would melt in summer. The period during which the ground is exposed to the sunshine would lengthen, the ground would absorb more warmth and, therefore, the air over it would be warmer. The effect of the first mild winter would continue through the summer and into the following winter and, probably fairly quickly, the area of permanent snow would decrease. There are several ways in which such a process might be initiated (see page 134), but a general warming would certainly be more likely if there were no continental land masses in high latitudes. The high-latitude air masses would be maritime and more strongly influenced by the heat-retaining capacity of the oceans.

There may be other effects. When a land mass breaks into two sections, for example, a new seaway is opened. This may allow currents to convey warm or cold water into formerly inaccessible sea areas. Just as the formation of air masses is altered, so is the circulation of ocean currents.

The geographical distribution of the continents and oceans determines the characteristics of the air masses and, therefore, of the general circulation of the atmosphere. Now that we know the continents have not always been in the positions they occupy today, and that they are still moving, we can amuse ourselves by imagining the climates of a world in which the continents were arranged very differently. What would happen, for example, were we to tow Africa and Eurasia southwards, so the Equator ran just north of the Himalayas and, at the same time, to tow Antarctica northwards into the Pacific — where there is room for it — so it lay across the Equator?

A map of the world would then show the Equator passing mainly over land, rather than sea, as it does now. There would be less land in high latitudes and none at the poles. The atmospheric circulation in the tropics would, presumably, continue to be dominated by the Hadley cells, but its air would be much drier. The tropics would probably be semi-arid and deserts would still cover much of the subtropics. If polar ice caps existed, they would be much smaller than those of today and, consequently, polar regions would be warmer. The temperature gradient between the Equator and poles would be shallower, the jet streams would be weaker — or even absent — and probably the climates in temperate latitudes would be more settled. Weather of a particular kind — wet or dry, warm or cool — would remain constant for longer than it does at present and might end with a spell of fairly violent weather — storms and gales — as one type of weather gave way to another.

This is pure speculation, of course, and could be quite wrong, but it serves to illustrate the great climatic significance of the distribution of land and sea at different latitudes. It is easy to see that while land lies in the tropics it will experience a tropical climate and that this can explain the presence of coal. Similarly, the desert which supplied the sand of what is now Devon probably lay in the subtropics. (The sandstone is now red because metals forming part of its mineral structure were oxidized when the sand was exposed to a more humid climate.) It is less obvious that a large change in the area of land near the Equator or near either pole also produces climatic effects which are felt throughout the world.

Successive ice ages

We cannot draw a map of the world as it was more than 2000 million years ago, but there is some reason for supposing that large land areas were located in high latitudes. There is evidence that parts of what are now North America, South Africa and Australia were covered by ice sheets about 2300 million years ago. Much later, from about 950 million years ago until about 615 million years ago, parts of what are now Africa, Australia and Europe lay beneath the ice. It is probable, but not certain, that North Africa was covered by ice about 450 million years ago.

South Africa, South America, India and Australia were partly covered by ice for a long time, between about 350 million years ago and about 250 million years ago, though not continuously. This episode was at the end of the Carboniferous and commencement of the Permian Periods, and it was at about the time that Pangaea began to break into two supercontinents. Both were moving northwards, but Laurasia drifted more rapidly and the Tethys Sea inundated the land that, until then, had joined them. During the Permian, the climates of the northern hemisphere were mainly continental — dry, with extremes of temperature — and ice was widespread in the southern hemisphere.

It is tempting to suppose that the movement of continents was responsible for these large-scale changes in the climates of the world. Continental wandering did not necessarily trigger the changes, but it would certainly have predisposed the world to them. There are other 'triggers' (see page 134), but pulling a trigger will not fire a shot unless the gun is loaded, and the location of continents represents the loading of the gun. Just as coal swamps occur only in low latitudes, so ice sheets are associated with high latitudes and the presence or absence of land near the poles makes ice sheets more or less likely to form.

A period during which part of the Earth is covered by snow and ice which do not melt during the summer constitutes an 'ice age'. Technically, therefore, we are living in an ice age now. So far as anyone can tell, the great Carboniferous-Permian ice age that occurred around 290 million years ago was the last for a very long time. For most of its history, the Earth has been free from year-round ice and ice has covered the southern hemisphere more frequently than the northern.

A new ice age

About 2 million years ago, or possibly a little earlier — with the continents more or less in their present positions — the world entered a new ice age. The period prior to that — the Pliocene Epoch — had been generally warm but the Pleistocene Epoch (the name is from the Greek words *pleistos*, meaning 'most' and *kainos*, meaning 'recent') brought dramatic changes. These led to an increase in the size of the Antarctic ice sheet — or its formation, because it may not have existed during the Pliocene. Apart from Antarctica, however, there is little land at high latitudes in the southern hemisphere and so the changes affected mainly the northern hemisphere. At its most extreme, ice covered about 28 per cent of the land area of the Earth — at present it covers 10.4 per cent.

Europe lay beneath the Scandinavian Ice Sheet, in places 10,000 feet (3000 m) thick, which extended as far south as Kiev in the Ukraine and covered Britain approximately to as far south as a line between the Severn Estuary and the Wash. Sea ice linked the Kamchatka Peninsula with the Aleutian Islands and Alaska. North America lay beneath the Laurentide Ice Sheet, extending from the Atlantic to the Rockies and as far south as the latitude of New York, Cincinnati and Kansas City. Over most of this area the ice was between 5000 and 10,000 feet (1500-3000 m) thick. Greenland and Iceland were almost completely covered by ice and sea ice extended across the Atlantic, linking Newfoundland with Ireland. There were also isolated ice sheets in various parts of Asia, western North America and western Europe, including a large sheet around the Alps.

The glaciers of Europe advanced during the Little Ice Age, but by only a short distance compared with their size during the ice ages. Apart from this episode, the ice sheets and glaciers have occupied their present positions throughout the whole of recorded human history, and the discovery that they had been much more extensive in the distant past came as a shock. Until then, the accepted view was that the world was essentially unchanging. Scientists knew of the existence of deposits which resembled glacial deposits, but were far removed from

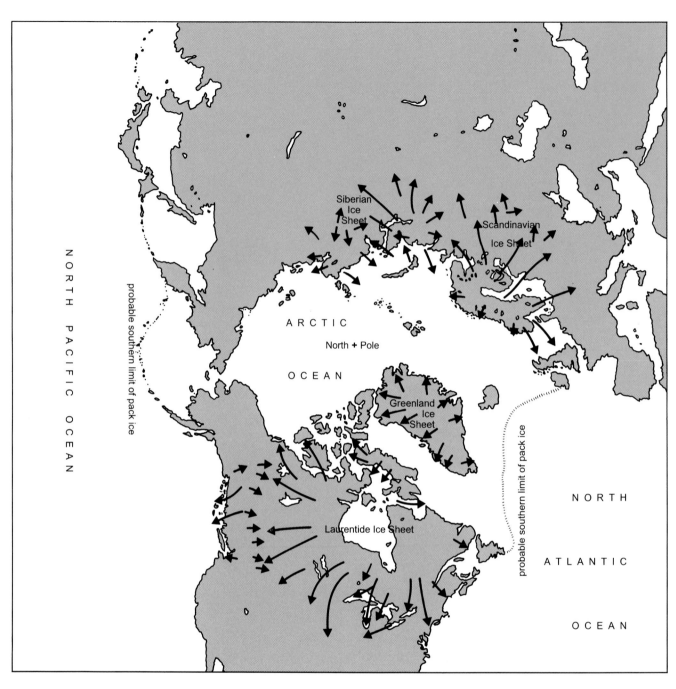

The northern part of the northern hemisphere showing the extent of the ice sheets during the Pleistocene Epoch.

the end of any glacier, but these were held to have been produced by the retreat of water, following the Biblical flood. This traditional view was overturned in the 1830s, principally by a Swiss zoologist, palaeontologist and geologist, Louis Agassiz (1807-73). Agassiz studied the deposits, first in Switzerland, and in 1837 he proposed that at one time — recently, in geological terms — the whole of Switzerland lay beneath an ice sheet just like the one covering Greenland. In 1840 he studied deposits in Scotland and concluded that Scotland had also been covered by ice. He moved to the United States in 1846, partly

to continue his studies and partly to deliver a series of public lectures on what by then was coming to be known as the 'Great Ice Age'. The lectures proved a sensation. In 1848, Agassiz was appointed Professor of Geology and Zoology at Harvard University and he remained in the United States for the rest of his life.

Once people knew it was there, the evidence of the 'Great Ice Age' could be seen to be abundant. The North American Great Lakes, for example, fill basins which were scoured by former ice sheets that advanced from the north. The glaciers pushed debris ahead of them and left vast amounts of it to the south of what are now the Lakes.

In Britain, too, there are many landscapes that were shaped by the ice. The long, deep, narrow lakes of the Lake District appear on a map like the spokes of a very irregular wheel. This is because they were scoured by glaciers that radiated outwards from a central ice cap.

Other evidence is more subtle. There are places in central Asia where patches of permanently frozen ground — permafrost — form 'islands' surrounded by warmer ground and a long way south of the more generally frozen ground near the Arctic Circle. These are areas that have still not thawed, remnants of conditions that were once more widespread. The permafrost layer is often hundreds of feet thick, but it is not necessarily easy to recognize, because it lies below a surface layer of ordinary, unfrozen soil.

Vegetation patterns can also remain long after the departure of the conditions they typify. In Upper Teesdale, in northern England, there are plant communities 'left behind' from the time when this part of Europe supported a tundra vegetation, of the kind found in northern Canada or Siberia. These relics of former times are called 'Pleistocene refugia', and some refugia are older even than that. On the Lizard Peninsula, in Cornwall, in south-western Britain, there are plants which are typical of a 'Lusitanian flora' that still remain from an interglacial that preceded the most recent advance of the ice.

Upper Teesdale, in northern England, contains plants typical of much colder conditions than those prevailing today.

Dartmoor, in south-west England, is a long way south of the polar ice sheet but its landscape is strongly reminiscent of the periglacial conditions found in much higher latitudes.

It is also possible for an entire landscape to survive as a reminder of a former climate. Much of Dartmoor, the famous upland moor in south-west England, is regarded by geographers as being 'periglacial' — occurring near the edge of an ice sheet. Parts of the moor are dotted with large granite boulders that look as though they were scattered by a giant. In fact, they were carried down gently sloping gradients during the period each summer when the permafrost melted and the ground surface turned to mud. The mud flowed, carrying the boulders with it, then froze again as winter returned. Today, Dartmoor is a very long way from the edge of the nearest ice sheet and there is no permafrost. The boulders have given up their summer migrations and, for thousands of years, they have lain as they were left when the permafrost thawed for

the last time.

The ice cover was not constant. Over the course of the 2 million years of the Pleistocene the edges of the ice sheets advanced and retreated many times. There were intervals, called 'interglacials', during which climates grew warmer — often warmer than they are today — and previously glaciated land was exposed and colonized by plants, animals and people. The ice never quite disappeared, however. Part of Greenland remained covered and there was permanent sea ice around the North Pole, in the middle of the Arctic Ocean. Antarctica remained covered by ice and surrounded by sea ice.

Retreat of the ice age
The ice sheets and glaciers began to retreat about 10,000 years ago. The dating of this

retreat and the concept of the 'Great Ice Age' led to the idea that the ice age had ended. The Pleistocene Epoch, the ice-age epoch, had come to an end and a new epoch had begun — called the Holocene, or Recent (the names are used interchangeably). Most geologists now believe the change of name may have been premature. There is still ice on the Earth, at both poles, and glaciers still exist. By definition, therefore, the ice age has not yet ended, and the discovery that during the 'Great Ice Age' the ice advanced and retreated several times leads, inexorably, to the conclusion that we are living in an interglacial — which also has a name, the Flandrian. If this interglacial resembles its predecessors, eventually it will end and the ice will advance once again. Indeed, the consensus among scientists who study the history of the ice ages is that the present interglacial is now very close to its end and a new ice age is due to commence at any time. To a geologist, 'any time' means within the next few thousand years, but starting now.

There were four major ice ages during the Pleistocene, separated by interglacials. All of them are named after the geographical regions in which the evidence of them was obtained, and so there is something of a profusion of names but also much uncertainty, because ice sheets do not advance and retreat at the same time and by the same amount on all their fronts. What is quite possibly the same event may have a North American, northern European, central European and, in some cases, a British name.

The earliest ice age is called the Nebraskan in North America and Donau in Europe. Then, from about 1.3 million years ago until about 900,000 years ago, the ice retreated. This interglacial is known as the Aftonian in North America, the Donau-Günz in central Europe and the Waalian in northern Europe. The interglacial gave way, about 900,000 years ago, to a less extremely cold ice age but one in which the ice advanced further in North America. This is called the Kansan in North America and the Günz in Europe. Then came the North American Yarmouthian interglacial, lasting from about 800,000 or 700,000 to 550,000 or 500,000 years ago, and known as the Günz-Mindel interglacial in central Europe and the Cromerian in northern Europe. At various times during this interglacial,

climates were warmer than those of today and at other times they were cooler. The North American Illinoian ice age, which followed, lasted from about 500,000 or 550,000 to about 300,000 or 400,000 years ago and is known in central Europe as the Mindel, in northern Europe as the Riss (or, by some authorities, Saalian) or Elsterian and in Britain as the Wolstonian. During this ice age, average temperatures were about 3.5-5.5 °F (2-3 °C) cooler than those of today. It was followed by the North American Sangamonian interglacial, lasting from about 550,000 to about 80,000 years ago, and equivalent to the Riss-Würm interglacial of central Europe, the Eemian of northern Europe and the Ipswichian in Britain (although it did not begin in northern Europe and Britain until about 100,000 years ago and ended about 70,000 years ago). It was during this interglacial that hippopotamuses and rhinoceroses lived in southern Britain. The most recent ice age, lasting from about 80,000 to 10,000 years ago, is called the Wisconsinian in North America, the Würm in central Europe, the Weichselian in northern Europe and the Devensian in Britain.

Ice sheets and glaciers mark landscapes in characteristic ways. Moraines consist of rock debris pushed ahead of moving ice, for example, and glaciers scour typically U-shaped valleys. Such features tell us nothing of what conditions were like during ice ages and interglacials, however. For that we must try to identify the plants and animals that lived at the time. Where the land surface has been little disturbed by humans, and soil and organic matter have been accumulating for a very long time, it is possible to remove cores of this material and extract pollen grains from it. Pollen grains are encased in an extremely tough covering which preserves them from decomposition, so they can be seen and studied under a microscope. Different plants produce pollen grains of a characteristic shape and, therefore, samples of pollen represent particular plants that can be identified. This makes it possible to reconstruct the vegetation that was growing at a particular place at a particular time, to produce, as it were, a 'snapshot' of the natural environment in the distant past.

What causes ice ages to begin and end

The general picture that emerges shows that, during an ice age, vegetational zones do not change but simply migrate to lower latitudes. Ice ages have little effect at the Equator so the vegetational zones become compressed in the temperate and high latitudes. A more interesting picture emerges as the ice retreats and an interglacial begins because then it is possible to trace the recolonization of areas by plants. The melting ice leaves behind a completely bare land surface, but the meltwaters feed

many rivers and, as the temperature rises, plants soon return and some species are very sensitive to small climatic differences and are useful indicators of conditions. The mountain avens (*Dryas octopetala*) (see page 122) grows naturally only in alpine regions or as a component of tundra vegetation, and holly (*Ilex aquifolium*) requires a mild, maritime climate. The recolonization of the British Isles since the end of the most recent (Devensian) glaciation is now fairly well documented.

During the most severe ice ages, the average temperature in the glaciated regions is 12.5-14.5 °F (7-8 °C) lower than it is at present, but a much smaller fluctuation — of as little as 5 °F (2.8 °C) — may be enough to set the ice advancing or retreating. The climates of the world are finely balanced and small changes can produce dramatic effects. It is difficult to anticipate such changes because, in the normal course of events, temperatures can vary widely from one year to another, and small variations in the averages can be detected only by examining a very large number of readings collected from widely separated places over a period that is long enough for a sustained increase or decrease to be apparent. Even then, the change may herald nothing more than a sequence of warm or cool years, a variation about a longer-term average. Usually, it takes at least 10 years for a fluctuation to become evident and about 20 years for it to be confirmed as a trend.

(Above right) The first step in reconstructing past vegetation patterns is to take a core sample, using an auger with a hollow centre that fills with soil as it is screwed into the ground. Withdrawing the auger calls for a strong back! (Right) Pollen grains, extracted from the sample and mounted on microscope slides, serve to identify the plant species. These are from hollyhock.

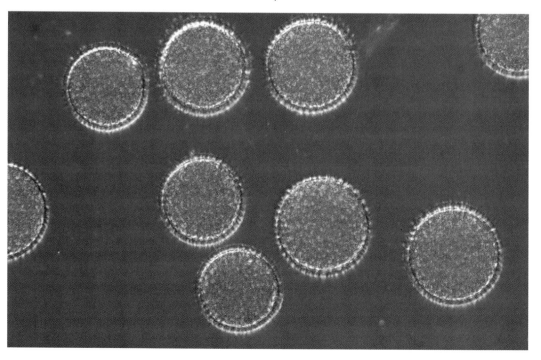

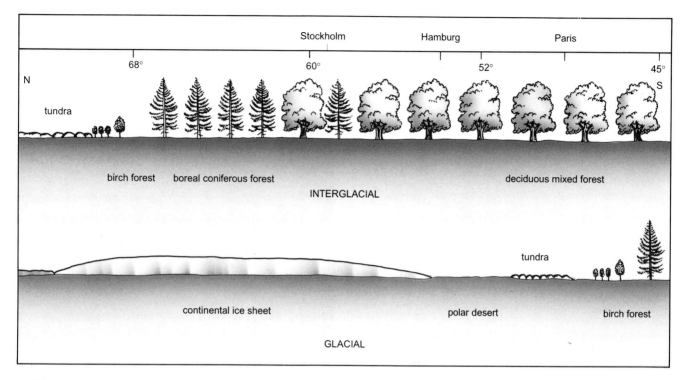

birch forest   boreal coniferous forest

deciduous mixed forest

INTERGLACIAL

tundra

continental ice sheet

polar desert

birch forest

GLACIAL

The change in the climate as ice ages begin and end affects more than the formation and character of air masses. The chemical composition of the atmosphere also changes. These changes 'force' the rate of climatic change — in both directions — so that, once the climate change begins, the chemical change accelerates it. Some scientists suspect that, in some cases, the chemical change may precede the climate change so it contributes to the change or even causes it.

There is clear evidence for the link between atmospheric composition and average temperatures. Analysis of air entrapped in the ice cores taken from the ice sheets in Greenland and Antarctica — from the Soviet Vostok station — has allowed scientists to reconstruct this link for the last 160,000 years — from the middle of the most recent glaciation. The chemical analysis of the air shows that each rise in temperature is accompanied by an increase in the concentration of carbon dioxide and methane in the air, and that each time the climate grows cooler, the concentration of these gases falls.

Changes in the chemical composition of the atmosphere force changes in climate and, to a large extent or, perhaps, completely, those chemical changes are the result of biological activity. The growth of green plants is based on the process of photosynthesis, in which carbon dioxide is

removed from the air and oxygen is returned to it. When the plants die, the carbon they contain is oxidized to carbon dioxide and, in the long term, the amount of carbon dioxide returned to the air is precisely equal to the amount that was removed. The plants and the atmosphere are in chemical equilibrium — but only while the total amount of plant growth, known technically as 'primary productivity', remains constant. If plant growth increases, more carbon dioxide is removed from the air than is returned to it, while a decrease in plant growth means more carbon dioxide is returned to the air by the decomposition of plant remains than is removed from it by photosynthesis.

A significant proportion of the photosynthesis takes place in the surface waters of the oceans where phytoplankton are abundant, and carbon dioxide is also removed by those marine organisms which use it to manufacture shells or plates. The coccolithophorids remove substantial amounts. Again, if their numbers increase or decrease, more or less carbon dioxide will be removed from the air.

Coccolithophorids, and other members of the plankton, also produce dimethyl sulphide (DMS), some of which oxidizes to sulphur dioxide. Sulphur dioxide provides cloud condensation nuclei and, therefore, an increase in the total mass of plankton will lead to increasing cloudiness. The removal of carbon dioxide will have a

The cold conditions of an ice age (glacial episode) compress the broad belts of natural vegetation types into lower latitudes, where the climate is milder. (Top) During an interglacial (as at present) deciduous mixed forest gives way to high-latitude (boreal) coniferous forest in about the latitude of southern Scandinavia, birch forest begins in about the latitude of central Finland, and tundra a little further north, about at the Arctic Circle. (Bottom) During the full Pleistocene glacials, the ice sheet extended as far south as Hamburg (and sometimes further), the belt of tundra gave way to birch forest south of the latitude of Paris, and the coniferous and deciduous forests began further south still.

The Soviet Antarctic research station, Vostok, where ice cores are taken to study the composition of the atmosphere in the distant past.

climatic cooling effect, but the associated increase in cloudiness is more complex. Clouds shade the surface and reflect incoming radiation. This has a cooling effect. The condensation that forms the clouds releases latent heat, however, which has a warming effect, and a warming of the atmosphere will increase the rate at which water evaporates. Water vapour is a 'greenhouse gas' (see page 161) and has a warming influence.

Living organisms exert considerable influence on climatic change, but purely physical processes are also at work. Starting in the late 1920s and with increasing intensity since the 1940s, scientists have been reconstructing the climatic history of northern Europe at the end of the last (Devensian or Weichselian) ice age. The eminent Danish botanist J. Iversen studied the pollen found at Bølling Sø and Allerød, in Denmark, and Sir Harry Godwin, who was Professor of Botany at Cambridge University, studied pollen from British sites. The work involved many other botanists and archaeologists — it is still continuing — and the picture that has emerged shows that the end of the last ice age was by no means a smooth, uninterrupted warming.

The warming began about 18,000 years ago and continued for about 5000 years. Then it was interrupted. Temperatures plummeted and, for about 1000 years, almost glacial conditions returned. Towards its end, between about 12,500 and 12,100

years ago, there seems to have been a short period of warming and cooling again before the general warming resumed. For the next 1000 years or so, the world grew warmer but then there was a second interruption and a second period, this time lasting for about 500 years, from 10,800 to 10,300 years ago, during which the climate was almost glacial. These sudden fluctuations — some scientists have calculated the climatic 'switches' may have taken no more than a few decades — are recorded in the pollen. During the colder episodes, the tundra plant *Dryas octopetala* (mountain avens) was common; during the warm interval it disappeared and birch (*Betula*) grew; then the birch vanished and *Dryas* reappeared. The two cold episodes are known as the Older Dryas and Younger Dryas, the warm interval separating them as the Allerød, and the brief warming at the end of the Older Dryas is called the Bølling, the Allerød and Bølling being named after their localities.

The fact of these climatic fluctuations was firmly established by the late 1950s, but the reason for them remained a mystery. It is only in recent years that measurements of the ratios of the two principal oxygen isotopes, O-16 and O-18, have allowed scientists to construct a narrative that is coming to be accepted as the most likely account of those remote events.

As the climate grew warmer, the part of the Scandinavian (also called the

Fennoscandian) ice sheet that extended over the sea between Norway and Denmark became unstable. It started to break apart, releasing icebergs which drifted southwards, melting as they did so. This chilled the surface waters and, because ice consists of fresh water which freezes at a higher temperature than salt water, there may have been drifting sea ice in middle latitudes, cooling the air crossing the Atlantic. The more significant effect, however, occurred further north, closer to the edge of the permanent sea ice in the northern Atlantic.

When ice forms on the surface of the sea, the salt dissolved in the sea water is not incorporated in the lattice of ice crystals and, therefore, the water adjacent to the ice becomes more saline. This makes the water denser. Its density is increased further by its low temperature, which is very close to freezing. Water is most dense at 39.2 °F (4 °C). This very cold, saline, dense water sinks deep beneath the surface to form a current flowing towards the Equator — the North Atlantic Deep Water. It is this cold current that drives the circulation of water in the North Atlantic, and the other aspect of that circulation is the northward transport of warm water by the Gulf Stream and the branch of it — the North Atlantic Drift — that breaks away to flow around the western coast of Britain and Norway.

Scientist believe that, as the ice sheet retreated and the melting of the ice released fresh water to float on the surface of the salt water (it floats because fresh water is less dense), the formation of North Atlantic Deep Water was inhibited and may have ceased entirely for a time. The driving mechanism for the oceanic circulation failed and warm water ceased to flow so far north — the North Atlantic Drift no longer broke away, and the Gulf Stream turned south in the latitude of southern Europe. This was probably sufficient to reverse the climatic warming — and it happened twice.

It is possible to explain how a climatic change may be forced once it has begun, and how a warming may be reversed, but these explanations cannot account for the onset or ending of an ice age. Ice ages are believed to be caused by reductions in the amount of radiation the Earth receives from the Sun.

Sunspots

The Little Ice Age may have been triggered

by just such a change. In 1889, a German astronomer, Gustav Spörer, drew attention to a period in which sunspot activity had been very low. The following year, the British astronomer E. Walter Maunder wrote a paper on the same subject and, in 1894, he published a paper called 'A Prolonged Sunspot Minimum'. In 1922, he published another paper, with the same title. Maunder was superintendent of the solar division of the Royal Greenwich Observatory in London but, despite his eminence, none of these attempts succeeded in arousing scientific interest in his discovery that the Sun is not constant — its output of energy changes from time to time.

The appearance and disappearance of dark spots (see page 155) on the surface of the Sun are the visible signs of this changing output. They have interested astronomers throughout history, although they were observed and recorded only by Asian observers until Galileo began recording them early in the seventeenth century. In 1843, a German astronomer, Heinrich Schwabe, discovered that the average number of sunspots changed very regularly. He calculated that the cycle takes 10 years, but it was found later to take 11.2 years. Maunder, however, checked the historical records and found that, although the regular cycle was more or less constant, its intensity was not — the number of sunspots appearing and disappearing over each 11.2-year cycle was not always the same. In particular, he found a period of 70 years, from 1645 to

Past climate in Britain Estimated average air temperatures (A) in the Lowland Range and (B) in the Highland Zone for the last 12,000 years. Lapse rates range between 5.6 °C (10.08 °F) and 7.6 °C (13.68 °F) per 1000 metres (3280 ft) as appropriate for 'Boreal' and 'Atlantic' phases, 1 Winter 2 Year 3 Summer

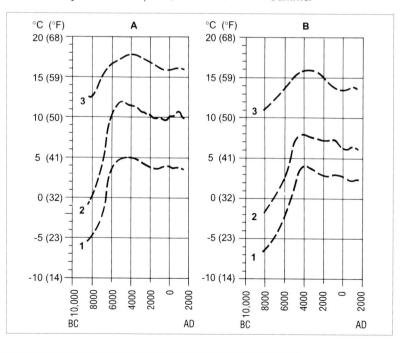

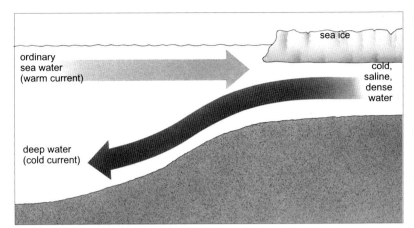

ordinary
sea water
(warm current)

sea ice

cold,
saline,
dense
water

deep water
(cold current)

(Opposite) The Earth's climate is influenced by periodic changes in the planet's rotation and in its orbit about the Sun. (Top) The present northern hemisphere seasons; the tilt of the Earth's axis inclines this hemisphere towards the Sun in summer, producing long days, and away from it in winter, producing short days. Bottom centre shows the angle of axial tilt, which changes over a period of about 41,000 years. (Middle) The Earth is furthest from the Sun (aphelion) on July 4 but, because there is a 'wobble' in the angle of the axis, the dates of aphelion and perihelion change over about 21,000 years. Bottom left shows how the 'wobble' causes the axis to describe a conical path. (Bottom right) Over about 95,000 years, the elongation of the Earth's elliptical orbit about the Sun changes. When the orbit is at its most elliptical, the Earth is further from the Sun at aphelion and perihelion, but closer at the equinoxes, than it is when the orbit is at its least elliptical.

1715, during which the total number of sunspots that were recorded was less than the number seen in an average year.

Maunder recognized that such a large change in sunspot activity might have profound implications on Earth, not least climatic implications. Yet his work remained obscure until, in the 1970s, John A. Eddy, a solar astronomer working at Harvard College Observatory and the Smithsonian Astrophysical Observatory, set out to check it. Eddy was intrigued, and he and his colleagues were thorough. They found evidence to corroborate what by then was known as the 'Maunder minimum' — it was a genuine phenomenon and not due to any failure in recording sunspots that were there but not observed. Then Eddy checked the implications of the minimum by comparing the number of sunspots recorded with climate records. The fit, in Eddy's words, 'is almost that of a key in a lock'. The Maunder minimum corresponds with the coldest period of the Little Ice Age. Eddy went further, and compared sunspot and climate records back to medieval times. Then he went beyond the reliable accounts of climate and used carbon-14 records, from which climatic conditions can be inferred. These allowed the study to be extended back to nearly 3000 BC and the close fit continued. Every increase and decrease in solar activity corresponds with climatic warming or cooling, and a change of as little as 1 per cent in solar output is enough to produce a significant change in average temperatures and the severity of winters.

Most scientists now accept that the Maunder minimum from 1645-1715 was linked to the Little Ice Age and that similar minima and maxima, with very small changes in solar output, influence the climate on Earth. In themselves they are

probably not enough, however, to trigger the really large-scale climatic changes associated with the onset and ending of major ice ages. Other mechanisms must be involved.

The most widely accepted theory of ice ages was proposed in the 1920s by a Yugoslavian physicist and mathematician, Milutin Milankovich (1879-1958). Milankovich drew together a number of contributing factors and expressed them mathematically on a graph as the 'Milankovich solar radiation curve'. The graph indicated nine periods, occurring at irregular intervals, when solar radiation was at a minimum and each minimum corresponded to an advance of the ice. It is not changes in the solar output that bring about these major climatic changes, but three types of variation in the Earth's orbit and rotation on its own axis. Each of these variations follows its own, very regular, cycle but each cycle has a different period. The mathematical task was to relate them.

The longest cycle affects the Earth's orbit itself. This describes not a circular path, but an elliptical one. The orbit is not quite regular, however. Gradually, the ellipse becomes more stretched, making it more elliptical, carrying the Earth farther away from the Sun at both ends of the ellipse and closer to the Sun in the middle part of the orbit. The stretching of the ellipse reaches a maximum and then it relaxes and the orbit becomes less elliptical. This is called the 'eccentricity' of the Earth's orbit and, starting from any point in the cycle, it takes about 95,000 years to return to the same point.

If you imagine a picture of the orbit of the Earth around the Sun with the space between the Earth and Sun filled in, the result will look like a flat surface — a plane. This is known as the 'plane of the ecliptic'. The Earth rotates about its own axis but the axis does not make a right angle with the plane of the ecliptic — the Earth is slightly tilted. It is this tilt which gives us our seasons as first one hemisphere and then the other is turned towards the Sun and then away from it. The angle of tilt is not constant. Measured against the plane of the ecliptic, it varies from a minimum of 21.8 ° to a maximum of 24.4 °, the complete cycle from maximum to maximum taking about 41,000 years.

Finally, there is the 'precession of the

equinoxes', which was first discovered by Hipparchus, the greatest astronomer of the ancient world, who was working around 130 BC in Rhodes and Alexandria (the dates of his birth and death are not known). You can demonstrate this effect for yourself, using a toy gyroscope. Set the gyroscope spinning and then nudge it, very gently, so its axis is tilted a little from the vertical — like the Earth's axis. Now watch what happens to the axis. You will see that it describes a circular path of its own. If you picture it extended some way above the top of the gyroscope the shape it makes will be a cone. This is what the precession of the Earth's equinoxes might look like to an outside observer, but it is not how it is caused. The fact was discovered by Hipparchus, but explained by Isaac Newton (1642-1727). It happens because the Earth is not precisely spherical. Technically, the Earth is an 'oblate spheroid'. In other words, it bulges at the Equator — each pole is about 13 miles (21 km) closer to the centre of the Earth than a point on the Equator. The gravitational attraction of the Moon and Sun produce a tidal effect on this additional mass at the Equator and the tidal force, combined with the rotation, causes the precession.

The tilt in the axis produces our seasons, but it is the precession of the equinoxes that determines when they occur in relation to the orbit around the Sun. At one extreme, the northern hemisphere winter occurs when the Earth is most distant from the Sun (aphelion), and at the other extreme it occurs when the Earth is closest to the Sun (perihelion). It takes about 21,000 years from one extreme through the cycle and back to the same extreme. About 11,000 years ago, perihelion occurred in June — the northern-hemisphere summer. At present, perihelion occurs in early January and aphelion in early July.

It is the interaction of these three cycles that Milankovich calculated and related to known dates of the advance and retreat of ice during the Pleistocene. The match was close enough to have convinced most climatologists that his solar radiation curve provides most of the explanation for the timing of ice ages and interglacials, although other factors — including Maunder maxima and minima — are likely to be involved.

The global climate is influenced by many factors — biological, chemical, physical and astronomical. Long-term predictions have to take account of all of them and of their interactions.

(Opposite) Formation of North Atlantic Deep Water. The formation of sea ice removes fresh water from the sea surface. Immediately below the ice the water is therefore more saline, and is at a temperature of about 39 °F (4 °C). These factors combine to make the water more dense than the water below it. It sinks below the less dense water, moving away from the ice edge, and forms the Deep Water flowing towards the Equator. Its place is taken by water flowing polewards. This mechanism is believed to drive the oceanic current systems.

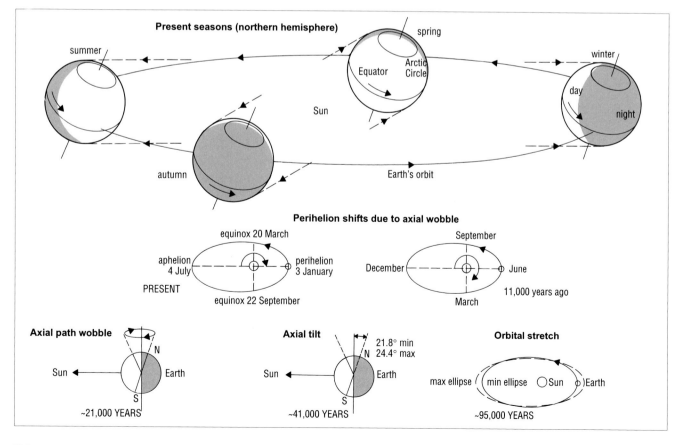

Present seasons (northern hemisphere)

summer   spring   winter

Arctic Circle   Equator   day   night

Sun   autumn   Earth's orbit

Perihelion shifts due to axial wobble

equinox 20 March   September

aphelion 4 July   perihelion 3 January   December — June

PRESENT   March   11,000 years ago

equinox 22 September

Axial path wobble   **Axial tilt**   **Orbital stretch**

Sun ←   N   Earth   Sun ←   21.8° min   24.4° max   N   Earth   max ellipse ((min ellipse ○Sun) Earth

S   S

~21,000 YEARS   ~41,000 YEARS   ~95,000 YEARS

THE RADIATION BALANCE

Daylight and the warmth we feel on a fine summer's day reach us as solar radiation. As everyone has known for many thousands of years, it is the Sun that makes life possible and that sustains it continually. The radiation we see and feel, however, represents only a small part of the total radiation emitted by the Sun.

The Sun is a 'main sequence' star. This sounds as though our local star is rather ordinary. It is, but that is not what the term means. It formed, as all stars form, when the particles comprising a cloud of dust and gas were attracted to one another gravitationally. They began to clump together and, eventually, the cloud collapsed. All the particles fell towards the gravitational centre. As they did so, they lost their gravitational energy. Energy cannot disappear, however. It can be neither created nor destroyed, but its form can change, so that one kind of energy is converted into another kind. In physics, this necessary conservation of energy is summarized as the first law of thermodynamics. As the falling particles 'lost' gravitational energy, that energy was converted into heat and the centre of the new star became hot. At the same time, the gravitational pressure near the centre became so strong, because of the large mass of material, that the nuclei of hydrogen atoms were forced to merge. This converted the hydrogen into helium with the release of energy.

This fusion of hydrogen atoms, with a release of energy, is called 'hydrogen burning'. Stars spend the greater part of their active 'lives' burning hydrogen which is why this is called their 'main sequence'. The Sun is a main sequence star because it is still burning hydrogen. So far, it has burned about half of its available stock — not all the hydrogen can be burned — and its present composition is about 90 per cent hydrogen, 8 per cent helium and 2 per cent consists of heavier elements.

The energy released by the fusion of hydrogen makes the centre of the Sun very hot — the temperature is calculated to be about 21.6 million °F (12 million °C). The high temperature causes a pressure to expand inside the core of the Sun, the gravitational pressure draws matter to the centre, and the two forces reach an equilibrium that determines the size of the Sun — its total diameter is about 870,000 miles (1,400,000 km).

Energy from the core of the Sun moves outwards but some of it is absorbed. Outside the core, there is a layer, about 310 miles (500 km) thick, called the 'convective zone', beyond that the 'photosphere', about 62 miles (100 km) thick and, finally, the 'chromosphere' which extends to about 9300 miles (15,000 km). These layers are heated by the radiation they absorb, and the surface temperature of the Sun is about 10,800 °F (6000 °C). Beyond the chromosphere, there is yet another layer more variable in depth, called the 'corona'. It is too faint to be seen except when the main part of the Sun is hidden from view — in a solar eclipse, for example — and it is produced by ionized atoms of iron, with a temperature of about 1,800,000 °F (1,000,000 °C).

The corona is losing matter, in the form of ionized atoms, protons and electrons, but not at a constant rate. From time to time, the rate of loss suddenly increases for a short time. This is a 'solar flare'. All the time, but with varying intensity, the particles streaming from the corona in all directions, at up to about 500 miles per second (800 km/s), form the solar wind. It is this wind that 'blows' the tails of comets into straight lines that are directed, not in the opposite direction to that in which the comet is moving like the smoke from a steam train, but away from the Sun.

The solar wind, consisting of particles, comprises only a small part of the total solar output, which is about 61.5 million watts for every square yard of the solar surface (73.5 MW per sq m). The Sun is losing matter in this way, but it is losing much more in the form of electromagnetic radiation. The light we see and the warmth we feel are both electromagnetic radiation.

Electromagnetic radiation
Electromagnetic radiation consists of a stream of photons. Photons are particles, but they are quite unlike other particles. They carry no electrical charge and it appears that they have no mass. You may wonder how it

(Opposite) Comet West, photographed in 1976. The tail, made from gas and minute dust grains, is 'blown' by the solar wind. It points directly away from the Sun, regardless of the direction in which the comet is moving.

At the National Center for Atmospheric Research at Boulder, Colorado, a research balloon is launched. The results from such balloons yield important information about the atmosphere and enable scientists to predict the effects that pollutants will have on future weather patterns.

is possible for a particle to exist and yet to have no mass. The answer calls for the suspension of our common-sense view of the world, which is necessary when considering the world of subatomic particles. A particle may have energy because it possesses mass and also because it is in motion. If a photon is stopped, however, it is destroyed and disappears, but it moves faster than any other particle. It is this that leads physicists to conclude that its energy consists only of the energy of its motion — it has no mass.

Continuing with this suspension of common sense, a photon (like other elementary particles) is also a wave. The wave extends over the whole distance it travels, but most of the wave characteristics are confined to a small region, and it is this

confinement that gives photons some of the properties of particles.

All photons travel at the same speed, the speed of light — in a vacuum, this is about 186,000 miles per second (300,000 km/s). Not all photons have the same amount of energy, however. Because a photon has no mass and it can move at only one, constant, speed, there is only one way in which the difference in energy between one photon and another can appear — in its wave characteristics. The greater the energy a photon has, the shorter will be the distance between one of its wave peaks and the next. Its energy can be expressed as its wavelength. Conversely, the closer together the waves, the greater the number of wave peaks that will pass a fixed point in a given time.

The number of wave peaks per second is called the 'frequency' of the wave and, therefore, the relationship between the wavelength and frequency of any regular wave motion is fixed, and the energy of the photon can also be expressed as its wave frequency. The wavelength is equal to the speed of propagation of the wave (in the case of electromagnetic radiation, the speed of light) divided by the frequency, and the frequency is equal to the speed divided by the wavelength.

Photons can be created. An atom consists of a nucleus surrounded by electrons. Each of the electrons possesses a particular amount of energy. This determines its distance from the nucleus, and it can change. When an electron moves to a lower energy state — it loses energy — a photon may be released. Changes can also occur within the nucleus of an atom and these may also release photons. The energies present in the nucleus are much greater than those outside the nucleus, among the electrons, and so a photon emitted from a nucleus has much more energy — at least 1000 times more — than one emitted from elsewhere in an atom. Photons emitted by atoms have relatively long wavelengths and low frequencies, therefore, and those emitted by nuclei have short wavelengths and high frequencies.

In the Sun, photons with very high energy — and short wavelength — are produced by the fusion of protons (hydrogen nuclei). As they move outwards, away from the core, some of them interact with atoms they encounter. These interactions lead to the emission of photons with less energy, and the end result is that the Sun emits photons with a vast range of energies, and physicists are able to associate particular photon energies with the region of the Sun from which they came.

The spectrum

The range of electromagnetic wavelengths is called its 'spectrum' and it is very wide indeed. The part of the solar spectrum of which we are aware — visible light and heat — is quite narrow, but it is the region in which the Sun radiates most intensely (see page 155). Because the spectrum is so large, different parts of it have been given their own common names. 'Visible light' is one such name and others include 'ultraviolet', 'heat', 'radio waves' and 'X-rays'.

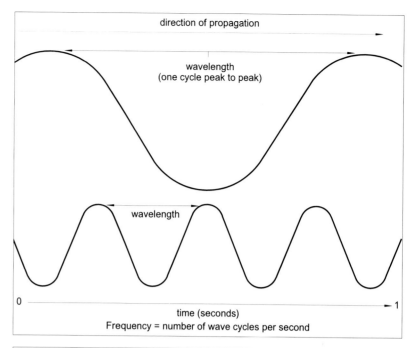

gamma 0.00000000001 — 0.0000001 mm	ionizing radiation
X-rays 0.00000001 — 0.000001 mm	
extreme UV 0.000005 — 0.0002 mm	
far UV 0.0002 — 0.0003 mm	
near UV 0.0003 — 0.0004 mm	
violet 0.00039 — 0.000425 mm	
indigo 0.000425 — 0.000445 mm	
blue 0.000445 — 0.0005 mm	visible light
green 0.0005 — 0.000575 mm	
yellow 0.000575 — 0.000585 mm	
orange 0.000585 — 0.00062 mm	
red 0.00062 — 0.00074 mm	heat
infra-red 0.00074 — 1.0 mm	
microwaves 0.1 — 20 mm	
radio waves 4.0 mm	radio and TV

overlap

1,000,000 mm +

(Above) Electromagnetic radiation can be imagined like waves on water. The distance between one wave peak (or trough) and the next is called the wavelength, the vertical distance between a peak and a trough the amplitude, and the number of peaks that pass a given point in one second is the frequency.
(Left) The solar spectrum (not to scale)

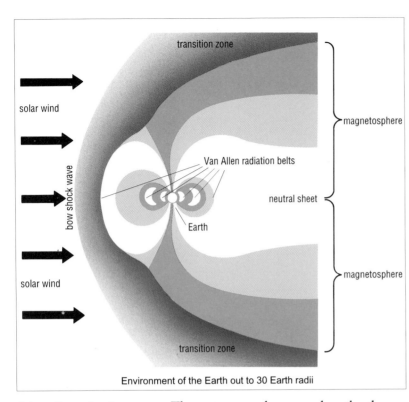

transition zone

solar wind

magnetosphere

bow shock wave

Van Allen radiation belts

neutral sheet

Earth

magnetosphere

solar wind

transition zone

Environment of the Earth out to 30 Earth radii

A two-dimensional representation of the three dimensional magnetic field (magnetosphere) that surrounds the Earth. Electrically charged particles streaming from the Sun (the solar wind) are trapped in the magnetic field and follow spiral paths along its lines of force. The pressure of the solar wind compresses the magnetosphere on the sunward side. In the outermost part (the boundary or transition zone) the magnetic field is relatively weak. Charged particles are most abundant in a concentric series of belts (Van Allen radiation belts), shaped like tyres which surround the Earth, located over the Equator.

There are several ways to describe the energy of a particular part of the electromagnetic spectrum, each with its own unit. The joule (J) is the scientific unit of energy and the electronvolt (eV) relates the energy of the photons to that of an electron. Conventionally written as one word, one eV is the the energy needed to move one electron through a potential difference of one volt. More commonly, the spectrum is divided into sections according to the wavelength or frequency of the radiation, always given in metric units — metres (m), millimetres (mm, thousandths of a metre), micrometres (μm, millionths of a metre) or nanometres (nm, billionths of a metre) — the unit being varied according to the length it describes. To compare one part of the spectrum with another, the simplest solution is to use only one unit, the millimetre (there are approximately 25.4 mm to 1 inch). The unit of frequency is the hertz (Hz), one Hz being equal to one wave cycle per second. This unit is often used to describe radio waves.

The most energetic end of the solar spectrum consists of radiation at wavelengths between about 0.00000000001 mm and 0.0000001 mm. This is the radiation emitted by events that occur inside atomic nuclei and its common name is 'gamma radiation'. When any atom is bombarded by electrons, it emits X-rays. These form the part of the spectrum with wavelengths of

0.00000001 mm to 0.000001 mm. Wavelengths of about 0.000005 mm to 0.0002 mm comprise the extreme ultraviolet (UV) part of the spectrum, between 0.0002 mm and 0.0003 mm lies the far UV and from 0.0003 mm to 0.0004 mm it is the near UV. The near UV section of the spectrum is sometimes subdivided further into UVA, UVB and UVC.

At wavelengths longer than those of UV, radiation becomes visible to our eyes, as light. Sunlight consists of a range of wavelengths each visible as a distinct colour, but which we see as white when they are mixed together. Literally, these are the colours of the rainbow, because the water droplets that break sunlight into the bands of a rainbow are separating it according to its wavelengths. The complete spectrum of visible light consists of violet (0.00039 mm to 0.000425 mm), indigo (0.000425 mm to 0.000445 mm), blue (0.000445 mm to 0.0005 mm), green (0.0005 mm to 0.000575 mm), yellow (0.000575 mm to 0.000585 mm), orange (0.000585 mm to 0.00062 mm) and red (0.00062 to 0.00074 mm). Sunlight is most intense at about 0.0005 mm, which is between blue and green light. Just beyond the end of the visible wavebands, infra-red radiation has wavelengths between about 0.00074 mm to 1 mm. We cannot see in the infra-red, but we can sense its longer wavelengths — they are what we feel as radiant heat. The waveband between about 0.1 mm and 20 mm, overlapping with the longer infra-red, comprises microwaves and the microwave band overlaps with the radio waveband. This begins at about 4 mm (very short wave) and extends over the remainder of the spectrum, to 1,000,000 mm or longer.

Protective role of the atmosphere

The Earth intercepts only about 0.002 per cent of the total amount of radiation emitted by the Sun. This amount is proportional to the distance between the Earth and the Sun and, at the average distance of about 93,200,000 miles (150,000,000 km), it comes to about 1131 watts per square yard (1353 W/sq m). This is the amount of radiation that would reach the surface were none of it deflected, reflected or absorbed on its way. It is called the 'solar constant', although variations in the output from the Sun mean it is not quite so constant as people once believed and, of course, it is an average — it increases as the Earth approaches perihelion and decreases as it approaches aphelion.

The figure for the solar constant seems low until you remember that the Sun radiates in all directions and the Earth is a very small target. Most of the solar output comes nowhere near our planet. Of the solar radiation that does reach us, the solar constant, about 51 per cent — an average of 595 watts per square yard (690 W/sq m) — penetrates all the way to the surface and about 19 per cent — 215 watts per square yard (257 W/sq m) — is absorbed in the atmosphere and so contributes to the total amount of energy available to us from the Sun.

The solar wind, which consists of particles, is deflected far above the surface, in the Earth's magnetosphere. As its name suggests, the magnetosphere is the outermost part of the Earth's magnetic field. As children, most of us played with a magnet and iron filings. Scatter the filings on to a sheet of paper, place the magnet beneath it, tap the sheet to make the filings move and they arrange themselves into a regular pattern. The pattern is that of the lines of force of the magnetic field, but this simple experiment is somewhat misleading, because it suggests that the field is confined to a plane — it is flat, or two-dimensional. A magnetic field is three-dimensional, of course, and extends more or less spherically from the magnet, the lines of force converging at the north and south magnetic poles. The magnetosphere has quite a definite boundary, but its shape is distorted and its size varies because of the pressure applied to it by the solar wind — which is not always constant.

Some of the particles that comprise the solar wind are electrically charged — protons carry a positive charge and electrons a negative charge — and charged particles are trapped by magnetic fields (the two forces, of electricity and magnetism, are really aspects of the same, electromagnetic, force). They move around the field, the several forces acting on them causing them to follow spiral paths along the lines of force.

The particles that carry no charge — neutrons — are also trapped by the magnetosphere, but at a lower level. They are able to penetrate the magnetic field but, as they approach the Earth, they collide with atoms of atmospheric gases. The collisions produce charged elementary particles, which are trapped in the magnetic field, and gamma-ray photons, which are not. The particles of the solar wind and the much smaller stream that arrives from other parts of the galaxy together make up the 'cosmic radiation'.

Incoming particles are prevented from reaching the surface, but photons are not

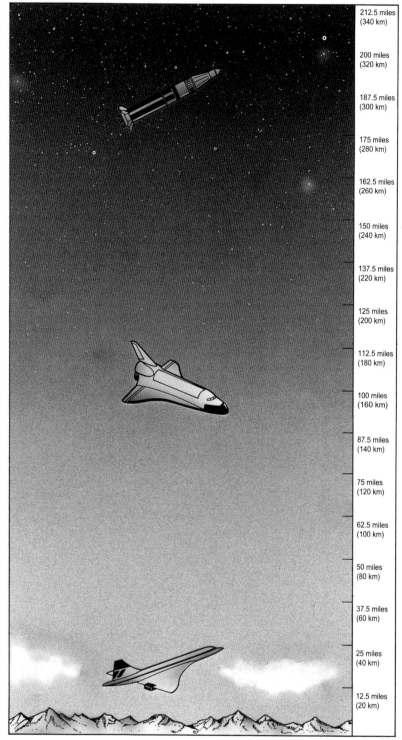

212.5 miles (340 km)

200 miles (320 km)

187.5 miles (300 km)

175 miles (280 km)

162.5 miles (260 km)

150 miles (240 km)

137.5 miles (220 km)

125 miles (200 km)

112.5 miles (180 km)

100 miles (160 km)

87.5 miles (140 km)

75 miles (120 km)

62.5 miles (100 km)

50 miles (80 km)

37.5 miles (60 km)

25 miles (40 km)

12.5 miles (20 km)

A schematic diagram of the atmosphere. High-energy radiation increases with altitude so that, to be able to explore the atmosphere above the levels of the highest mountains, different vehicles are needed to cope with the changing conditions.

affected by the magnetosphere. Their fate is determined by their wavelengths.

Gamma rays, X-rays and extreme UV collide with atmospheric atoms and molecules and lose much of their energy in the impact. The atoms may lose electrons, ionizing them and releasing free electrons. This accounts for the composition of the ionosphere. The photons continue on their way, but at longer wavelengths. At a lower level, radiation with wavelengths between about 0.0007 mm and 0.001 mm encounters a greater density of oxygen molecules and loses energy by splitting the molecules into single atoms, some of which reform in threes, as ozone. Ozone itself absorbs radiation between about 0.0003 mm and 0.0007 mm, in the UV waveband. Water vapour also absorbs incoming radiation, in three narrow wavebands around 0.0015 mm, 0.0019 mm and 0.0021 mm to 0.004 mm. Clouds and dust particles also absorb a small amount.

The overall effect of the atmosphere is to prevent any radiation with a wavelength of less than about 0.0003 mm from reaching the surface. Virtually all of this radiation is absorbed at heights above 6 miles (10 km).

Very short-wave radiation is able to penetrate many substances, passing between the molecules, before it collides with an atom or molecule and, when it does collide, it has sufficient energy to ionize atoms. An ionized atom carries a positive electrical charge and this makes it extremely reactive. Inside living cells, such highly reactive molecules are extremely disruptive. Short-wave radiation — gamma and X-rays — is called 'ionizing radiation' and is hazardous. UV radiation, with a longer wavelength, has no penetrating power, but it can break large molecules apart. It, too, is hazardous, but only mildly so.

Although the atmosphere affords ample protection to organisms at the surface, if you climb far above the surface, exposure increases. Assuming you have no protection — and a layer of ordinary clothes will provide full protection against UV — your exposure to near UV will increase significantly as you climb above the ozone layer (see page 159). As you rise still higher, the intensity of UV radiation will increase and its wavelength will shorten. If you climb as far as the ionosphere — to a height of, say, about 62 miles (100 km) — you will be exposed to significant amounts of X-rays

and gamma rays, against which clothing affords no protection.

In the lower atmosphere, molecules with a radius of less than one-tenth of the wavelength of radiation will scatter the radiation. Scattering of blue light by oxygen molecules accounts for the colour of the sky. Larger molecules and particles will scatter radiation of any wavelength. Scattering of all wavelengths by larger particles accounts for the whiteness of a hazy sky. A proportion of the scattered light is directed away from the Earth and so it is not absorbed.

Certain particles in the atmosphere, clouds, and the Earth's surface also reflect solar radiation, the amount depending on the albedo (see page 148).

Aurorae

Travel to northern Canada or Alaska, or cross the sea between Norway and Iceland and, on most clear, dark nights, you will be treated to the most spectacular sight the atmosphere can produce (see page 156). You have entered an approximately circular belt, about 300 miles (483 km) wide, called the 'auroral zone' — there is a similar zone in the southern hemisphere, in Antarctica. This is where aurorae are most often to be seen, and they quite often occur simultaneously in both hemispheres. If you are close to the auroral zone, but outside it, you are most likely to see an aurora in the direction of the zone, but they fade as you move closer to the poles or into lower latitudes. Near to their low-latitude limit you are most likely to see them in spring or autumn.

The sky will be illuminated by vast, draped curtains and arcs of light, usually white, yellow or green, with red near the edges. When aurorae become visible in lower latitudes, they are often much redder. There may be one 'curtain' or more. Sometimes there are as many as seven. Usually they appear to be waving gently, as though in a light wind, but they can move rapidly.

They are named after the Roman goddess of the dawn, Aurora and, because they occur close to the poles, the phenomenon is known technically as the 'aurora polaris' or, more commonly, as 'aurora borealis' or 'northern lights' in the northern hemisphere and 'aurora australis' or 'southern lights' in the southern hemisphere. Despite being named after a Roman goddess, aurorae are seen only very rarely in the Mediterranean region. Some Greek and Roman writers had heard of them, but few had witnessed them. At one time, people believed they were seeing fire, or the reflection of sunlight from snow. They are caused by the Sun, but they are neither sunlight nor fire. Captain James Cook (1728-79), the English navigator, was the first person to report seeing the aurora australis, in the Indian Ocean in 1773.

The apparent height of the 'curtains' depends on the position of the person seeing them. From a distance of about 700 miles (1126 km) they can seem to touch the horizon and, from still further away, they may seem to arise from below it. Their actual dimensions have been calculated trigonometrically by photographing them simultaneously from several different points. The base of the lights is between 50 and (more usually) 65 miles (80-105 km) above the surface and they extend upwards, commonly for about 100 miles (160 km), but sometimes much further, and rays from them can extend for several hundred miles above the upper edges. If the 'drapes' are moving slowly they may be about 2 miles (3 km) thick but, if they are moving rapidly, they will be much thinner.

The auroral zones are approximately circular, slightly elliptical belts with an average radius of between 20° and 25° of latitude. The zone in the northern hemisphere describes a circle centred in north-west Greenland, not far from Thule. The southern-hemisphere zone is centred on a corresponding point in Antarctica about 78.6 °S and 109.9 °E, in Wilkes Land, not far from the Soviet Vostok station.

These centres are hundreds of miles from the geographical North and South Poles, and they are also hundreds of miles from the magnetic poles — the northern magnetic pole is over north-western Canada. In fact, they mark the location of the 'geomagnetic poles'.

The magnetic field of the Earth — the 'geomagnetic field' — is of the kind that would be produced if, in the centre of the Earth, there were a dipole magnet. That is an ordinary magnet, like a bar magnet, with a north pole at one end and a south pole at the other. The core of the Earth is not shaped like a bar, of course, and does not really contain a bar magnet, but that is the shape of its magnetic field. The theoretical dipole — the 'bar' — is not aligned with the rotational axis of the Earth, however, but is inclined to it at an angle of about 11.3 °.

Draw a line through the centre of the Earth along the line of the dipole and the points where the line breaks the surface are the two geomagnetic poles.

A compass will not help you find either of the geomagnetic poles. This is because rocks between the core of the Earth and the surface respond differently to the geomagnetic field and, therefore, distort it. A compass needle will identify a magnetic pole. It follows magnetic field lines, which run from one pole to the other. The poles themselves are in the core of the Earth and, therefore, when the needle is directly above the pole it points either vertically downwards, at the north magnetic pole, or directly upwards, at the south magnetic pole. These are known as the 'dip poles' — because the compass needle dips — and, because of the distortions in the magnetic field, there are many of them and their positions wander. This is of little importance for most people, but when it matters — in navigating ships and aircraft, for example — magnetic compass readings must be corrected and the data used to correct them must be updated regularly. Away from the surface of the Earth, the geomagnetic field becomes regular and relates directly to the geomagnetic poles, not to the 'compass' poles.

Cosmic radiation, principally the solar wind, showers the Earth with protons and electrons. As they enter the geomagnetic field, the associated electrical field accelerates them, so they gain energy, although scientists are not yet certain why this should happen. Then, trapped in the magnetic field, they follow the lines of force in the field, spiralling around them and oscillating between the north and south geomagnetic poles. The lines of force and, therefore, the path of the electrically charged particles, sweep out far from the Earth over the Equator, but descend to the surface at the geomagnetic poles. A constant stream of charged particles is being drawn downwards at these points in both hemispheres. The stream descends from all directions, as though the particles were being drawn down a funnel or cone. This is why the auroral zones are approximately circular in shape and, although the 'curtains' may appear to rise vertically, in fact they are slightly inclined.

As they descend through the outer atmosphere, the particles encounter the atoms and molecules of the atmospheric gases — at this height, principally atomic oxygen and atomic nitrogen. The atmosphere is very tenuous and an incoming particle can travel an average of 6 miles (9.7 km) before it encounters one, but the deeper it penetrates, the more closely packed the atmospheric atoms become — at the surface, a particle can travel an average of only about 0.000003 inches (0.00008 mm) before colliding with one. Sooner or later, therefore, collisions are inevitable.

Aurorae are produced more by electrons than by protons. When an electron collides with an atom it may dislodge an electron from the atom, and it also transfers some of its own surplus energy to the electrons around the atomic nucleus. They become 'excited', but very quickly return to their former energy levels and, as they do so, the atom emits radiation — photons. This is called 'excitation radiation' and its wavelength is characteristic for atoms of each element. Some of the protons, meanwhile, capture free electrons. An electron that is bound to a proton has less energy than a free electron and so this, too, causes an electron to fall to a lower energy level, again with the emission of radiation. In this case, the wavelength is characteristic of hydrogen because the nucleus of a hydrogen atom consists of a single proton, and one proton with one electron form a hydrogen atom.

Aurorae have been produced artificially, thereby confirming the reason they occur. In the days when atomic bombs were tested in the atmosphere, a few were exploded at very high altitudes over the Pacific Ocean and the consequences monitored. Electrons flew upwards, were trapped by the geomagnetic field, and descended over the geomagnetic pole in the opposite hemisphere. The aurorae they produced were visible in lower latitudes than most natural aurorae, because the electrons had more energy than those arriving from space and the response was more intense. Whole aurorae have also been observed from above, by satellites, which confirmed their shape.

Sometimes, however, the upper regions of aurorae are visible from much lower latitudes because the aurorae extend to much greater heights. They have often been seen in Europe, occasionally even as far south as the Mediterranean, and in Mexico. These rare events occur during periods of increased solar activity. The solar wind intensifies — increasing the density of the particle stream — during sunspot maxima and decreases during sunspot minima. This causes particles to collide at a greater altitude.

Some of the electromagnetic radiation the Earth receives from the Sun is absorbed, some is reflected and some is scattered. Over long periods, the climates of the Earth remain fairly constant, and changes of just a degree or two in average temperatures bring dramatic changes. It follows, therefore, that the amount of energy leaving the Earth must equal the amount arriving.

Maintaining the balance

Incoming and outgoing energy can be compared like the credit and debit columns of a balance sheet. A calculation of this kind produces a 'radiation budget' which, like a financial budget, must balance. It can also be expressed as a 'heat budget'. Units of radiant energy can be translated into heat units because all the incoming energy that is absorbed — by the atmosphere as well as at the surface — is converted into heat. 'Absorption' means that energy is delivered to the atoms and molecules of the receiving substance, raising their temperatures. Even photochemical processes cause energy to be released as heat or, in the case of photosynthesis, to be stored for later release as heat.

The balances do not have to be achieved immediately or locally. There can be, and are, imbalances according to season and latitude. In high latitudes, for example, the ground surface loses more energy over the year than it receives and, in low latitudes, where the Sun is more directly overhead, the ground receives more energy than it loses.

There can also be very local variations. A hillside that faces away from the Equator may never be warmed by direct sunshine and may lose more energy than it receives, while the other, Equator-facing, side of the same hill receives more than it loses. Elsewhere, there are seasonal imbalances, with gains exceeding losses during the summer and losses exceeding gains during the winter. Even in a particular place and during a particular season, the balance can change from one hour to the next when clouds shade the surface. It is only over a number of years and over the planet as a whole that the totals must balance if the world is not to become hotter or colder.

The Sun behaves as an almost perfect 'black body'. That is to say, it radiates energy very efficiently, at an intensity related to its surface temperature of about 11,000 °F (6000 °C). If the Earth were also an efficient black body, promptly radiating back into space all of the energy it receives, the average surface temperature on the planet would be about -9 °F (-23 °C). The Earth would be locked into a permanent ice age more severe than any it has ever experienced. Clearly, this is not the case.

The wavelength at which a black body radiates is determined by its temperature, which is exactly what you would expect. The wavelength of radiation represents the amount of energy it conveys, and so the higher the temperature of the radiating body, the more energetic the radiation it emits will be.

The average temperature on the surface of the Earth — over the whole surface and a full year — is about 59 °F (15 °C). At this temperature, the Earth radiates very slightly at less than 0.004 mm, but mainly between 0.004 mm and about 0.04 mm, and most intensely at around 0.01 mm. The short-wave limit, of 0.004 mm, is taken as the boundary between 'short-wave' and 'long-wave' radiation. Where the temperature is higher — over the subtropical deserts, for example — the wavelength is towards the shorter end of the waveband, and where the temperature is lower — over high latitudes — it is towards the longer end.

Where more or less radiation is absorbed, the intensity — measured as the wavelength — at which the surface radiates will alter correspondingly. This will maintain a balance, but at a higher or lower surface temperature than the average. During the subtropical summer, for example, the wavelength of black-body radiation becomes shorter and the surface temperature rises. It is differences in the amount of radiation absorbed in different latitudes that produce our climates, as heat is transferred from warmer to cooler regions through the atmosphere and oceans.

The amount of energy involved is considerable, at least by human standards. A commonplace local shower of rain involves about the same amount of energy as is required to keep a modern airliner airborne for 24 hours, and a local thunderstorm releases as much energy as the burn-

ing of 7000 tons of coal. 'Powering' the Asian monsoon requires as much energy as was used in 1950 by all the factories, offices, cars and homes in the world.

Of every 100 units of short-wave solar radiation that reach the top of the atmosphere, an average of about 45 reach the surface and are absorbed. Approximately 31 of the incoming units are reflected back into space by clouds (23), the surface (4) or atmospheric particles (4). This is the planetary albedo (see page 148). About 4 units are absorbed in the stratosphere, mainly by ozone, 17 are absorbed in the troposphere, mainly by water vapour, and 3 are absorbed by clouds.

The Earth emits long-wave radiation but this does not leave the planet immediately. Atmospheric carbon dioxide, water vapour and ozone absorb almost all radiation at wavelengths below 0.008 mm and 0.012 mm. At about 0.01 mm, however, which is where the Earth radiates most intensely, there are gaps — 'windows' — between the absorption bands of water vapour and ozone and between those of ozone and carbon dioxide at about 0.015 mm. Especially when the air is fairly dry, long-wave radiation escapes through these windows.

The radiation that is absorbed in the atmosphere, together with latent heat released by the condensation of water vapour, is also re-radiated in the same way and in all directions. Some escapes directly into space, some is directed downwards and some is directed to the sides, where it is absorbed and radiated again or scattered in all directions by atmospheric molecules and particles. It is because of the absorption of long-wave radiation in the atmosphere — the 'greenhouse effect' — that the average temperature on Earth is held above its theoretical black-body value.

Of each 100 units of energy arriving at the top of the atmosphere, 31 are reflected. The atmosphere absorbs 24 units and a further 107 units of long-wave radiation from below, 23 units in the form of latent heat and 6 units of radiant heat. It loses 63 units of short-wave radiation to space and 97 units of long-wave radiation to the surface. The surface absorbs 45 units of short-wave radiation and 97 units of long-wave radiation, and loses 113 units as long-wave radiation, 23 units as latent heat, and 6 units as radiant heat. This balances the budget overall.

Albedo

The word 'albedo' is derived from the Latin *albus*, meaning 'white' and, in a sense, it is a numerical measure of whiteness or, more correctly, of the proportion of radiation a surface reflects in the short-wave band between 0.0003 mm and 0.004 mm. In summer, many of us wear white clothes, or at least light-coloured ones and, in winter, we tend to favour darker colours. We believe that white clothes reflect heat and dark clothes absorb it. This is true, although the number of layers of clothes and the materials from which they are made are more important factors in keeping us warm or cool than their colour.

The Earth as a whole reflects about 31 per cent of the solar radiation falling upon it. This figure, the Earth's total reflectiveness, is its 'albedo' value. It can be expressed as a percentage (31per cent) or as a decimal fraction (0.31). It is important, because any change in this value represents an increase or decrease in the amount of solar energy that is absorbed. The climatic implications are obvious.

The albedo of the whole Earth is, obviously, a composite of different albedos for the many different surfaces the planet presents to the Sun. Each type of surface has its own albedo value. The whole Earth has about the same albedo as sand, but this figure includes the albedo of the atmosphere and its clouds. The surface of the Earth has an overall albedo ranging from about 0.14 to 0.16.

Within regions, albedos change from time to time. Snow cover in winter increases albedo and the spring thaw reduces it again. Clouds form and clear. In the case of large expanses of water, the albedo depends on the wind and on the position of the Sun above the horizon. If you watch the Sun setting over the sea or a large lake, the light reflected from the water may dazzle you — the albedo is very high — but only if the water is calm. Waves will reduce the glare greatly, because they present different angles to the Sun. At midday, however, the water appears dark, because the Sun is higher in the sky. The angle between the radiation and the water surface is larger and much of the radiation penetrates — water is fairly transparent to radiation in the visible-light waveband — and is absorbed, principally by plankton and particles suspended in the water. The difference is very large. The albedo of water surfaces ranges from 0.06 for rough water to 0.20 for calm water, but is much higher when the Sun is low.

As you would expect, fresh snow is the surface with the highest albedo. Depending on how smooth and clean it is, its value lies between 0.8 and 0.9 — between 80 and 90 per cent of the radiation falling upon it is reflected. As the snow melts, its albedo falls to between 0.4 and 0.6. Sand comes next in the ranking, with an albedo of 0.30 to 0.35 — desert sands tending towards the higher value. Grassland and cereal crops on farms — which, botanically, are grasses — have an albedo of 0.18 to 0.25. Cereal crops are harvested, of course and, when the fields are ploughed, the albedo, assuming the ground surface is dry, falls to between 0.12 and 0.20. Wet soil is darker still, with an albedo of about 0.10. This helps the farmer, because it means the ground absorbs more warmth when it is bare and this assists seed germination.

The albedo of forest depends on the type of forest. Deciduous forests, of the kind that grow naturally in temperate regions with seasonal climates, have an albedo of 0.15 to 0.18. Coniferous forest is darker, with an albedo of 0.09 to 0.15, and tropical rain forest is a little darker still. Its albedo is between 0.07 and 0.15.

Bare rocks have an albedo of 0.12 to 0.18. The albedo of urban areas is the same as that of rocks. Buildings and roads are made from rock, after all.

The colour of a surface has such a strong influence on the proportion of solar radiation that is reflected and absorbed that it is a major factor in determining the climatic conditions a region enjoys. In fine weather, winter holidaymakers often protect their eyes and skin against the glare of light reflected from the snow, but the ground temperature remains low. The snow and ice do not melt and the air is not warmed by its contact with them. The solar radiation is often intense, but almost all of it is reflected back into space and lost.

The climatic consequences of altering the albedo are among the factors that concern scientists when large areas of land are converted from one use to another. Such changes often involve increasing albedo. When temperate forest is cleared for pasture, for example, the albedo of the surface increases from, say, about 0.16 to about 0.21 — although, if the land is ploughed, it

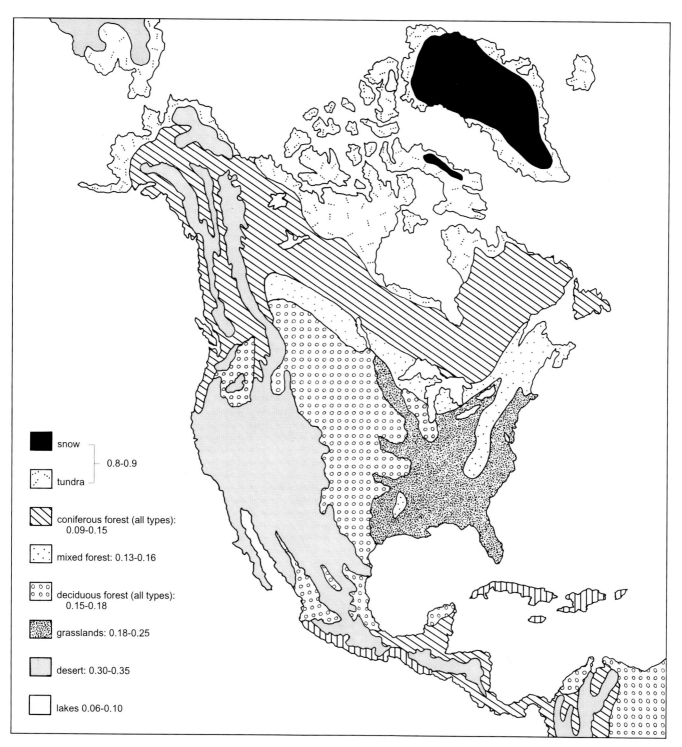

snow ⎤
             ⎬ 0.8-0.9
tundra ⎦

coniferous forest (all types): 0.09-0.15

mixed forest: 0.13-0.16

deciduous forest (all types): 0.15-0.18

grasslands: 0.18-0.25

desert: 0.30-0.35

lakes 0.06-0.10

changes little, or even falls while the soil remains bare. The conversion of tropical rainforest to pasture has an even larger effect, increasing the albedo from about 0.11 to 0.21, and the deterioration of grassland into sandy desert involves a still greater increase, from 0.21 to about 0.33.

The effects of clouds and precipitation
The effects on the climate are complex. An increase in its albedo means a surface

absorbs less radiation. This will lower its temperature and that of the air above it which, in turn, will reduce the rate of evaporation of water and cloud formation. Less cloud, however, will partly compensate by allowing more radiation to reach the surface.

If you have ever flown above the cloud tops you will know how brightly they shine in the sunlight. Stand on the ground as a large, threatening storm cloud drifts over-

The albedo of the whole Earth is a composite of the different albedos for the different surfaces which it presents to the Sun. This is a map of North America to show the variation in albedo from the many different types of natural surfaces.

Seen from an aircraft, it is obvious that the brilliantly white tops of cumulus cloud reflect most of the sunlight falling on them.

head, on the other hand, and it may grow so dark that people turn on the lights in nearby houses. The presence or absence of clouds makes a very great difference to the amount of radiation reaching the surface.

Water vapour and droplets absorb radiation but, compared with the albedo of cloud, the amount is insignificant. The proportion of sunlight that reaches the

surface when the sky is cloudy depends on the extent of the cloud cover, but also on the type and thickness of the clouds themselves. Estimates of the extent of cloud cover are included in all reports from meteorological stations. The figure is usually expressed in 'oktas' — one okta is equal to one-eighth of the sky — but sometimes you may see it reported in tenths or as a percentage. The effect on the local radiation budget is calculated by deducting from the incoming radiation an amount determined by the type of cloud and the number of oktas.

When you look at clouds from the ground or from an aircraft they appear solid. You know they are not, of course, but they appear to have surfaces, distinct boundaries that are clearly visible. It is natural, therefore, to assume that light is reflected from these surfaces and, indeed, this is the impression they give. The cloud tops look much like snow-covered ground. As your aircraft enters the cloud, the illusion of solidity is destroyed, but not the illusion of their defined surfaces. This is because aircraft usually enter clouds at high speed, so the interval is very brief between being outside the cloud and inside it. Once inside, visibility is reduced greatly — because, beyond the wing tips, there is nothing to see.

Clouds are really much less substantial than they look. Aircraft fly through them in close formation with no difficulty because, although visibility is reduced, it is not reduced sufficiently to reduce the clarity with which each pilot can see all the other aircraft in the formation. Nor do clouds have sharp boundaries. Fly at about the height of the tops and the aircraft will enter and leave cloud repeatedly and the cloud itself will be wispy, like a thin mist. People who walk in hilly country often encounter mists of this kind and, in fact, such hill mists are low cloud. They obliterate the panoramic view of the landscape but, during the middle part of the day, they are rarely so dense as to make it difficult to discern the path. Visibility is more severely reduced around dawn and sunset, because the Sun is lower in the sky and, therefore, its more oblique radiation must pass through a greater thickness of mist and more of it is lost.

It is evident, therefore, that, despite their reflective brilliance, clouds are not opaque. Light can and does penetrate them and inside a cloud, but near the top, the light can be very bright. Pilots are taught that, when flying in cloud, they must rely wholly on their instruments to judge the attitude of the aircraft, because their ordinary senses become very unreliable. They may suggest that the aircraft has one wing much lower than the other, or even that it is almost upside down, and correcting for such misjudgements is likely to prove catastrophic. The disorientation occurs, at least partly, because we expect sunlight to reach us from above and the sky to be lighter than the ground. This may not be so inside a cloud. It may be brighter below than above — suggesting the aircraft is inverted — because light is being reflected upwards from water droplets deep inside the cloud.

As it penetrates a cloud, the incoming radiation encounters a diffuse mass of water droplets. When photons strike droplets they are reflected or scattered. Imagine you are looking through a cylindrical column that passes right through the cloud, parallel to the incoming radiation so that, from your point of view, the thickness of the cloud is reduced to a disc. The degree to which the disc is covered by water droplets — the total area of droplets as a proportion of the area of the disc — is the 'column density' of the droplets in the cloud, and the distance photons can travel before colliding with a droplet depends on the column density. The column density depends on the depth of the cloud.

A cloud that is, say, 70 feet (21 m) thick will reflect about 40 per cent of the incoming radiation (its albedo is 0.40). This is about the albedo of cirrus but most clouds are much thicker. A cloud 200 feet (61 m) thick has an albedo of about 0.65, one 400 feet (122 m) thick 0.78 and one 600 feet (183 m) thick 0.80. Very deep clouds, such as cumulonimbus storm clouds, may extend for 10,000 feet (3000 m) or more and have an albedo of 0.90.

Such high albedo values suggest that cloud cover cools the ground below, but matters are not quite so simple. Apart from warming the surrounding air by absorbing a small amount of incoming radiation and by the latent heat of condensation, water vapour absorbs long-wave radiation between about 0.006 mm and 0.007 mm, and between about 0.022 mm and 0.05 mm. This radiation, from the surface, is prevented from escaping into space and so clouds have a warming as well as a cooling effect.

During the day, when the intensity of incoming radiation exceeds that of outgoing radiation, clouds block incoming radiation and have a cooling effect. At night, when incoming radiation ceases, they block outgoing radiation and so have a warming effect. Their overall influence is to make hot days cooler and cold nights warmer.

THREATS TO STABILITY

There are some years when the British summer is cool, wet and not much different from winter — and 1845 was just such a year. Conditions were ideal for the proliferation of a fungus, *Phytophthora infestans*, and, although no one knew it at the time, this was the cause of late blight of potatoes. The blight devastated the crop throughout the British Isles. The following summer was hot and sultry, with a drought that ended in June in Ireland and in August in England. The late blight destroyed the potato crop for a second time and in Ireland, where potatoes were a main staple food, there was famine. By 1851, one-eighth of the Irish population had died and more than a million people had emigrated.

There was one place that escaped the blight, however. Down-wind from a copper smelter in Wales the potatoes grew normally. Had scientists known the cause of the blight they might have realized the implication — smoke from the smelter contained copper, and copper poisons fungi. As it was, Bordeaux mixture, the first copper-based fungicide, was not discovered until 1878, by the French scientist, P.M.A. Millardet. A mixture of copper sulphate and lime, it stained grapes blue and had been used for some time to discourage theft from vineyards. Millardet noticed that treated vines were not attacked by mildew — another fungus.

A pollutant is usually defined as a substance released into the environment as a result of human activity and that either does not occur naturally, or does not occur at the concentrations which result from its release. Whether or not they are harmful depends less on their intrinsic toxicity than on the dose to which living organisms are exposed and the duration of their exposure.

Pollutants alter the chemical or physical character of the local environment but the consequences are difficult to predict, even today, and depend to a large extent on our cultural perceptions of advantage and disadvantage. The emissions from the Welsh copper smelter were beneficial, but this was quite fortuitous. Had people relied on mushrooms rather than potatoes, the weather of 1845 and 1846 would have produced bumper crops, except down-wind of that smelter, where they would have been destroyed.

We are inclined to think of industrial pollution as a recent phenomenon, but its history is long. The smelters that refined metals for imperial Rome filled the air with smoke — and in the ancient world there were no controls on emissions from metalworking. When Sodom and Gomorrah were destroyed, 'the smoke of the country went up as the smoke of a furnace' (Genesis 19: 28).

Air pollution

People were complaining about the quality of London air as long ago as the twelfth century and, by the sixteenth century, with the burning of coal well established, the air was so bad that, in 1578, Queen Elizabeth stayed out of the city to avoid the smell. The coal burned at that time contained much more sulphur than the coal burned today. In his entry for 11 April 1656, the diarist John Evelyn described a new technique for processing coal to remove its

Potato blight, caused by the fungus *Phytophthora infestans*, causes the leaves to wither and fall and prevents any further growth. The tubers will be small and inedible.

sulphur and arsenic, with coke as the end product. During the seventeenth century, there were allegations that Londoners were being poisoned by industrial air pollution. Other cities were no better. The air was filled with smoke in Sheffield, Newcastle-upon-Tyne, Oxford and, indeed, in most large towns. The fumes from the manufacture of bricks were considered especially noxious, and pollution from the industry was debated in Parliament in 1657. A few years later, the Duke of Chandos, who lived in Cavendish Square in what is now London's West End, complained of being poisoned by the smell from factories in general and the brick kilns in particular.

Industrial air pollution is as ancient as the use of fire to smelt metals and fire pottery, and for centuries people have known that it was harmful to health. Diseases were believed to be caused by 'miasmas' — noxious emanations, usually recognized by their bad smells — and smoke that makes eyes smart and sets people coughing is likely to be injurious, after all. There were many victims of respiratory disorders to supply proof, if proof were needed.

From about 1760, the rapid industrializa-tion of the European and American econo-mies during the Industrial Revolution exacerbated the situation. New industrial towns grew up and added to the forests of smokestacks and the volume of their dis-charges, until the smoke and industrial haze that had once covered cities spread to blanket entire regions. There was no longer any room for doubt about the harmful consequences. Municipal buildings and statues were blackened with soot. Many were made from limestone and were badly eroded as the calcium carbonate reacted with airborne sulphuric acid.

The human cost began to be counted and the more serious pollution incidents documented. In 1930, for example, air pollution killed 60 people and made hun-dreds ill in the Meuse Valley, in France. In 1948, in Donora, Pennsylvania, 18 people were killed by air pollution and nearly 6000 were made ill. In 1950, a release of hydrogen sulphide killed 22 people and made 320 ill at Poza Rica, Mexico. Then, in 1952, for four days London endured one of its notori-ous 'peasouper' fogs — fog made from water droplets mixed with soot and smoke and known in Britain since 1905 as 'smog'. During those few days, 4000 people died —

An industrial landscape of smoking factory chimneys. These are at a steel works at Duisberg, in the Ruhr, Germany.

(Above) Photochemical smog in Mexico City. (Right) Acid pollution damage on Regency buildings in Cheltenham, England. The buildings are made from locally quarried limestone.

twice the number of deaths that occur during the same period in most years. Exactly 10 years later a second London 'smog', lasting five days, killed 700 people. These episodes led to the passing of the Clean Air Acts of 1956 and 1968. British smogs — 'pea-soupers' were not confined to London — no longer reduce visibility to a few feet, causing street lights to be turned on at noon and workers to be sent home early, gasping for each painful breath. In the United States, the Air Quality Act was passed in 1967 and its provisions strengthened by the Clean Air Act, 1970.

Burning carbon-based fuels

The word 'smog' is now more commonly applied to a different form of air pollution. When nitrogen oxides and unburned hydrocarbons are trapped in humid air beneath a temperature inversion and exposed to intensely bright sunshine, photochemical reactions occur that produce a range of products including peroxyacetyl nitrate (PAN) and ozone. Photochemical smog is rare in Britain because the intensity of sunlight is usually too low, but it is common in lower latitudes wherever there is a high concentration of motor vehicles, which are the source of the nitrogen oxides and hydrocarbons.

The main source of air pollution is still the burning of carbon-based fuels. The emissions depend on the composition of the fuel and the temperature and efficiency of its combustion. Heavy fuel oil and coal contain varying amounts of sulphur which may be released as sulphur dioxide when the fuel is burned. Combustion at high temperatures — most notably in modern, high-compression internal combustion engines — produces oxides of nitrogen. A shortage of oxygen during combustion may cause incomplete oxidation of the fuel and the emission of carbon monoxide. Once mixed with air, this oxidizes rapidly to carbon dioxide. Carbon monoxide levels often rise in the vicinity of heavily congested, slow-moving traffic, though not to concentrations that are harmful to humans, but people have been killed by carbon monoxide that has accumulated in a confined space due to inefficient combustion. Inefficient combustion also releases a range of hydrocarbon compounds, many of which are harmful and some of which are suspected of

Acid rain can cause thinning of the crowns of many coniferous tree species. This may inhibit their growth, make them more vulnerable to infections and pest infestations and, in extreme cases, kill them. These trees are in the Erz Gebirge (Ore Mountains), Germany.

Emissions enter the air as gases or solid particles. Some gases react and particles are produced. Particles may fall to the surface or be washed down by rain. Gases may contact surfaces of buildings or vegetation, or may dissolve in rain.

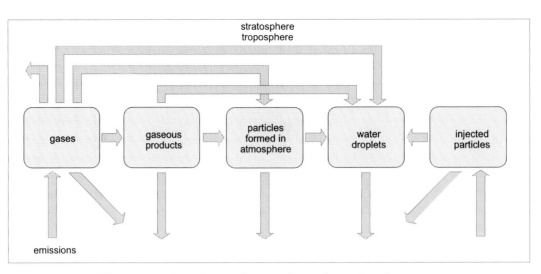

(Opposite, top) Sunlight, which we see as white, is composed of radiation of different wavelengths. When a prism separates these wavelengths we see them as bands of different colours. They are the colours of the rainbow because water droplets, acting as prisms, split light in the same way. (Opposite, bottom) The appearance and disappearance of sunspots, appearing as dark patches on the surface of the Sun, are related to changes in the intensity of solar radiation and the solar wind.

causing cancer. The principal products of hydrocarbon combustion, however, are carbon dioxide and water vapour. Traditionally, these have not been considered pollutants, because they are harmless to humans and occur naturally in the air. Carbon dioxide is now regarded with more suspicion, however (see page 163).

Other pollutants may be carried into the air with the hot combustion gases. Fine particles of ash, for example, may contain a range of potentially harmful substances, and lead, added to gasoline as tetraethyl lead to improve the anti-knock quality of the fuel, is suspected of accumulating to harmful levels in the bodies of children who live in busy urban areas. As awareness of the hazards of air pollution has grown, the introduction of improved technologies has increased the efficiency with which fuel and industrial materials are used and this has reduced emissions, although ample room remains for further improvement.

The incidental release of pollutants in the form of emissions is only one source of air pollution, however. There are also accidental releases — and deliberate ones. Accidents can be very serious. In December, 1984, the release of methyl isocyanate following the accident at a pesticide factory in Bhopal, India, killed about 2500 people and injured 200,000. Advances in industrial practices can reduce the likelihood of accidents, but it can never eliminate entirely the possibility of them.

The spraying of pesticides and the use of chlorofluorocarbon (CFC) aerosol propellants are examples of the deliberate release into the environment of substances that are classed as harmful years after they came into use. They illustrate more clearly than

industrial or vehicular emissions — although the point is the same — how difficult it is to predict the long-term consequences of any modification of the natural environment.

Acid rain

At one of their meetings in 1852, members of the Manchester Literary and Philosophical Society listened to a paper presented by a local scientist, R.A. Smith, 'On the air and rain of Manchester'. Smith had found that close to Manchester the rain was very acid and that its acidity decreased the further from Manchester he went. 'Acid rain' was certainly not a new phenomenon, but Smith may well have been the first person to notice and measure it. The following year, scientists at Rothamsted Experimental Station, in southern England, began to analyse the rainwater they collected but, apparently, it was not unduly acid then — although it is very likely to have been acid over more heavily industrialized regions. Systematic monitoring over the whole of western Europe did not begin until the 1950s.

It was in the 1960s that Swedish scientists associated severe corrosion of buildings in Stockholm with high sulphur concentrations in the air and, soon afterwards, they began collecting evidence of the increasing acidification of lakes in southern and central Sweden. Norwegian and Finnish scientists recorded similar observations and, in the early 1970s, abnormally acidic rain was reported in the north-eastern United States and southern Canada. During the 1980s, acid rain was reported to be causing serious damage to lakes and forests in central and eastern Europe.

All rain is naturally acid. The air contains carbon dioxide and oxides of nitrogen, which are slightly soluble in water and, therefore, dissolve into the water droplets in clouds to form dilute carbonic and nitric acids. Sulphate crystals act as condensation nuclei and the water droplets into which they dissolve become dilute sulphuric acid. Acidity is measured on a 'pH' scale — 'pH' means 'potential of hydrogen' — in which 7 represents neutrality, values lower than 7 acidity and values higher than 7 alkalinity. 'Clean' rain and snow have a pH of about 5, but in regions affected by 'acid rain', the pH is markedly lower than 5.

The cause is complex and differs from place to place. In Scandinavia and North America, acidity is due mainly to sulphuric acid from dissolved sulphate. Sulphur dioxide, released from the burning of coal and heavy oils — mainly in power stations — is an important source, and the industrial sulphur dioxide which affects Scandinavia comes principally from Sweden, the USSR, Poland, Czechoslovakia, Germany and Britain. A significant, but variable, amount is produced by the oxidation of dimethyl sulphide released by marine algae in the North Sea. German forests are not affected by sulphur dioxide — lichens, which are very intolerant of sulphur, grow abundantly — but have been injured by nitric acid and ozone, produced mainly by vehicles, combined with drought and disease. Vegetation is affected less by rain, which tends to drain away quickly, than by mist and the direct deposition of sulphate crystals from dry air. Coniferous trees tend to concentrate acidic moisture.

When acid precipitation reaches the surface, its effect depends on the local geology. In limestone areas, bicarbonate already present in the water will buffer it, allowing considerable quantities to be absorbed with no increase in the acidity of the receiving water. Where local rocks are predominantly igneous — as in southern Scandinavia — there is little or no buffering and the waters become increasingly acidic.

Plant leaves may be injured directly by acid deposition and the uptake of nutrients by roots can be affected by the chemical changes it produces in the soil — principally the 'locking up' of magnesium, which can be remedied by adding magnesium fertilizer. Many aquatic animals are very sensitive to changes in the acidity of the water in which

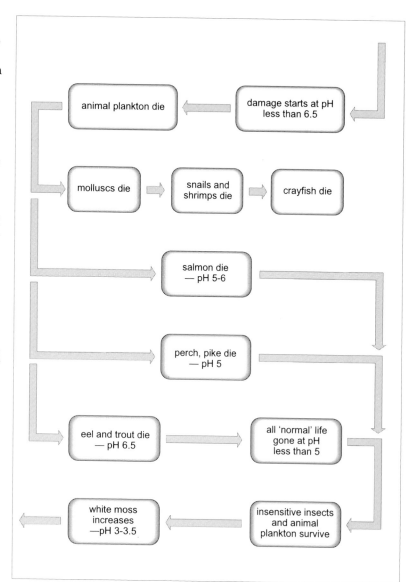

they live. The effects begin to appear — with the death of plankton and then various invertebrates — when the pH falls below 6.5. As the pH falls below 6.0 most fish are killed.

The ozone layer

Near the stratopause, at a height of about 30 miles (50 km), solar radiation at the short-wave end of the ultraviolet (UV) waveband splits oxygen molecules into their constituent atoms and the UV energy is absorbed by the atoms. The atoms rejoin, some of them in threes to form ozone, and ozone also absorbs UV radiation, its three molecules separating. The repeated breaking and reforming of oxygen and ozone molecules block virtually all UV radiation at wavelengths below about 0.0004 mm. Mixing of air in the stratosphere carries some upper air, enriched in ozone, to lower levels where

Aquatic animals are sensitive to the acidity of the water in which they live. As acidity increases, the most sensitive species are lost first and, if acidification continues, others follow. Monitoring aquatic animal species provides a guide to changes in the acidity of the water. (Opposite, top) Aurorae are caused by the bombardment of the upper atmosphere by the particles of the solar wind. (Opposite, bottom) The familiar shimmer, seen close to the ground on a hot, sunny day, is caused by the expansion and rise of air that is heated by contact with the ground.

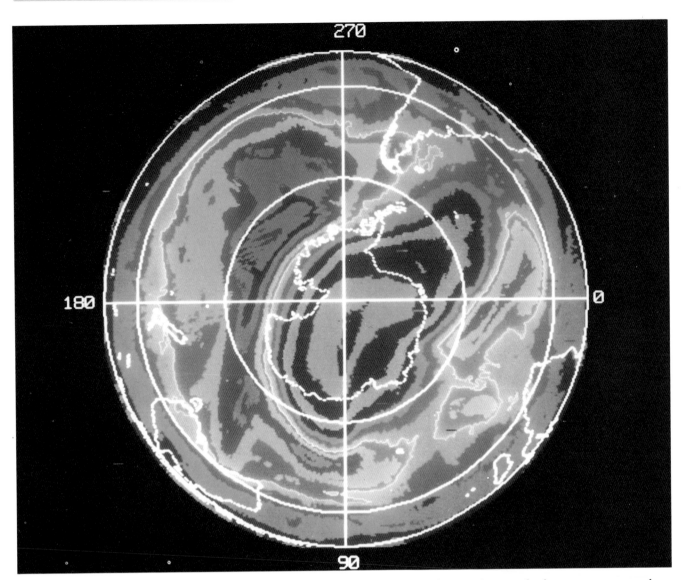

The photograph of Antarctica, taken on 10 October 1986, by Total Ozone Monitoring System (TOMS) on the Nimbus-7 satellite, which revealed the existence of the 'ozone hole' covering virtually the entire continent. 'Dobson units' measure the total concentration of ozone.

the ozone is below the main area of UV absorption and accumulates, especially between about 12 and 16 miles (about 20-25 km). This is the ozone layer.

Ozone is extremely reactive and, therefore, it is destroyed rapidly at lower levels, where it is more likely to encounter compounds it can oxidize. Above the ozone layer, ozone is destroyed by radiation and below the ozone layer it is destroyed chemically — although it is also formed in the troposphere by photo-chemical reactions.

There is little exchange of gases or particles across the tropopause but, since the 1960s, there have been fears of a gradual accumulation in the stratosphere of substances that could react with ozone. Were this to happen, ozone might be broken down faster than UV radiation causes it to form, and there would be a proportional increase in the amount of short-wave UV radiation reaching the surface. In 1987,

evidence of ozone depletion was reported from Antarctica — the so-called 'ozone hole'.

During the long Antarctic winter, stratospheric air moves in an approximately circular path, with a region of very still air — a vortex — at the centre. The temperature in the vortex, about -130 °F (-90 °C), is low enough to cause water vapour — probably produced by the oxidation of methane — to form polar stratospheric clouds made from ice crystals. Various nitrogen compounds adhere to surfaces of the ice crystals and, as the winter draws to an end and the returning Sun supplies a little energy, these compounds react with free chlorine. This, in turn, forms a stable oxide with ozone, depleting the ozone layer. As spring advances, the air circulation changes, breaking down the vortex. Air containing ozone is drawn in and, with the increasing intensity of the sunlight, the

formation of ozone is resumed.

Depletions have also been reported over the Arctic. These are much smaller, however, and last for only a week or two, because the winter vortex is much weaker — it does not form every year — and temperatures are usually too high for the formation of polar stratospheric clouds.

Chlorine is the key ingredient in the polar depletion of ozone. More than half of the chlorine that enters the stratosphere arrives in the form of methyl chloride (CH_3Cl). This is released when surface vegetation is burned — in bush and forest fires, for example — and by many wood-rotting fungi. It is chemically stable but, once in the stratosphere, UV radiation breaks the bond linking the methyl (CH_3) and chlorine (Cl) ions, liberating free chlorine.

The chlorine from methyl chloride provides a natural background that changes little over the years. The amount of stratospheric chlorine has increased mainly because of the use of chlorofluorocarbon compounds — CFCs. These compounds, of which there are several, are chemically very stable and do not react with the ordinary constituents of the atmosphere. Some CFC molecules are washed to the ground by rain, some enter the oceans and some adhere to solid particles, such as sand grains, but a proportion remain airborne and seep across the tropopause into the stratosphere.

The ozone layer is thickest in high latitudes in summer. Were it to disappear over middle latitudes altogether, the intensity of UV exposure at the surface in the latitude of Britain and Seattle would be about the same as that at the Equator. Current predictions, of a possible 10 per cent depletion within 50 to 75 years, would increase UV exposure by the equivalent of moving about 100 miles (160 km) towards the Equator — say from London to Dieppe or Seattle to Portland. There could be a small reduction in yields of sensitive crops and sunbathing could lead to more rapid burning, but the biological consequences are unlikely to be severe. Monitoring suggests, however, that in recent years the intensity of UV radiation over the continents has been decreasing rather than increasing.

The 'greenhouse' effect

Long-wave radiation — 'black-body' radiation — is absorbed in the atmosphere by a range of gases (see page 147) but escapes through 'radiation windows'. This temporary retention of heat by the atmosphere is called the 'greenhouse effect', and it serves to keep the Earth much warmer than it would be otherwise. If the atmospheric concentration of the 'greenhouse gases' increases, however, the existing radiation windows may be partly closed. Provided there were no compensating effect, this would lead to a warming of the atmosphere.

Polar stratospheric clouds to the north of Stavanger, Norway, at a height of almost 39,000 feet (11,887 m). The lower, orange band is of 'Type I' cloud, made from nitrogen trihydrates. The upper, 'Type II' cloud is made mostly of ice crystals.

(Above) Some scientists predict more hurricanes as a consequence of greenhouse warming (see page 103 for an explanation of how hurricanes occur). Hurricane-force winds cause great damage. These huts, in Guam, were destroyed by typhoon Roy in 1988. (Right) The rate at which tropical forests are being cleared has caused alarm throughout the world. This clearing is in the Amazon Basin.

A warming might also result from any increase in the intensity of incoming solar radiation. Stratospheric absorption in the ultra-violet waveband reduces the energy that can penetrate to lower levels and, therefore, an increase in UV penetration, resulting from ozone depletion, is likely to cause a cooling of the stratosphere and a

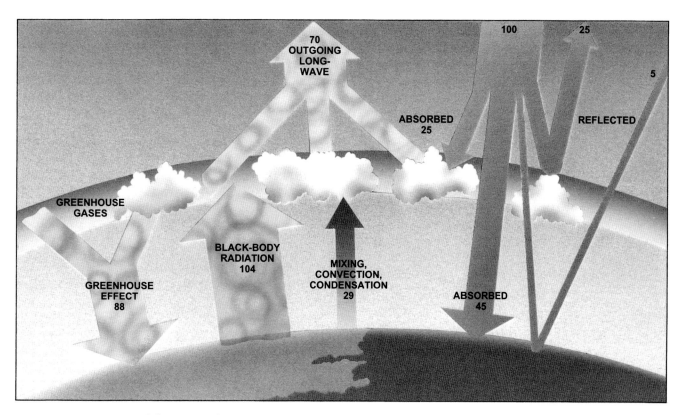

consequent warming of the troposphere.

Almost any gas molecule that consists of three or more atoms is likely to be of a size comparable to the wavelength of part of the long-wave electromagnetic spectrum, and such a molecule will absorb some outgoing radiation. Not all molecules are of equal importance. It may be, for example, that a particular gas is already so abundant that it absorbs all radiation in its waveband. Adding more of that gas will have no effect but adding another gas, even in small amounts, may cause radiation to be absorbed in a part of the spectrum that previously was open.

Carbon dioxide is the best known of the greenhouse gases, but this is only because it is the most abundant. It contributes about 50 per cent of our present greenhouse effect. Methane contributes about 18 per cent, CFCs 14 per cent, tropospheric ozone 12 per cent and nitrous oxide 6 per cent. Concentrations of all of them are increasing.

Over the past century, the proportion of carbon dioxide in the air has increased from about 290 to 350 parts per million. It is still rising by about 1.5 per cent a year, mainly because of the burning of carbon-based fuels and the clearing and burning of forests. Methane, released mainly from cattle and the growing of paddy rice, is increasing at about 1 per cent a year. Nitrous oxide is

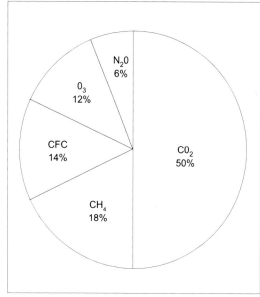

released when organic substances are burned, and ozone is produced by photochemical reactions involving nitrogen oxides and unburned hydrocarbons.

The average global temperature today is about 1 °F (0.5 °C) higher than it was a century ago. This may or may not be due to a greenhouse warming and, beyond predicting a general rise in temperature, calculating the climatic implications of the warming in each region of the world is very difficult. Most scientists agree it would be wise to reduce emissions of greenhouse gases.

(Above) The greenhouse effect. Of each 100 units of solar radiation the Earth receives from the Sun, about 30 are reflected by clouds or the surface and about 70 are absorbed and then re-radiated, leaving the Earth as long-wave radiation. The escape into space of long-wave radiation is delayed by greenhouse gases, which absorb most of it and re-radiate it in all directions so that it 'leaks' into space a little at a time, although the overall radiation budget is in balance. Because of this greenhouse effect, the global climate is much warmer than it would be otherwise. Should the greenhouse effect increase, delaying still more the escape of long-wave radiation, average temperatures are likely to increase in proportion.
(Above left) Greenhouse gases — percentage contribution to the greenhouse effect.

PRESERVING STABILITY

The Worldwatch Institute, based in Washington DC, published a report early in 1990 in which it estimated that one-fifth of all the people in the world regularly breathe air in which the pollution levels are above the limits recommended by the World Health Organization. The report said that the effect of breathing the air in Bombay, for example, is equivalent to smoking 10 cigarettes a day. Possibly the most widespread urban air pollutant, sulphur dioxide, is corrosive to respiratory tissues as well as to limestone. In 1989, the 'top ten' cities in terms of sulphur dioxide levels were, in descending order of contamination, Milan, Shenyang, Xian, Beijing, Rio de Janeiro, São Paulo, Seoul, Tehran, Paris and Madrid. From time to time, pollution levels in other cities exceed the recommended maxima — they did so in London and Dublin in November, 1989.

The 'pea-souper' smogs no longer occur in British industrial cities and a generation of children has grown up with no experience of them, but air pollution remains a serious problem. There is little doubt that, at least in some places, breathing urban air is injurious. People who must spend much of their working time outdoors have sometimes been advised to wear 'smog masks', which filter out solid particles but are probably ineffectual against gases, and at times it has been considered necessary to close schools temporarily for the protection of the children. It has been reported that, during particularly severe episodes of pollution, the death rate in Athens rises sixfold.

Such pollution is confined to urban areas, of course, for the atmosphere has a great capacity for self-cleansing, but it can be carried into neighbouring and otherwise unpolluted regions. 'Transfrontier pollution' drifts across national boundaries — and can be regulated only by international agreement.

Particles seldom remain airborne for more than a few hours or, at most, days. If they are relatively large, they fall gravitationally and settle or adhere to surfaces with which they come into contact. Smaller particles may act as condensation nuclei for water vapour or dissolve into or adhere to water droplets and be washed from the air by rain. Oxygen is also a powerful cleanser because it oxidizes many particles and gas molecules. Oxidation usually renders a compound less reactive and decreasing its reactivity makes it less hazardous — chemical pollutants cause damage by reacting with living cells or tissues.

Controlling pollution

There is nothing new about our pollution of the air and, over the centuries, many attempts have been made to control it. What was probably the first smoke abatement law was passed in England in 1273, during the reign of Edward I and, in 1306, Edward tried again, by issuing a proclamation forbidding the burning of coal in London. He was determined to succeed — one offending workshop owner was beheaded. Pollution from the factories of the Industrial Revolution called for a new and more comprehensive approach. In 1874, the first smoke abatement law in the United States was passed, in Chicago, and in 1880 the British Government established an 'Alkali Inspectorate', charged with reducing emissions from 'alkali and other works'.

In succeeding years, most countries passed legislation to deal locally or nationally with different types and sources of pollution, and various broad, underlying principles evolved. Many European countries, and now the European Community, calculated the dose of each pollutant humans or other species can tolerate, allowed a large safety margin, and then used this to recommend or impose limits to discharges. In the United States, legislation is often based on a defined satisfactory quality for air within a particular region — 'ambient air quality standards' — and local emission limits are set at levels likely to maintain these standards. Britain avoided the setting of precise limits, relying instead on the greatest possible reduction of all harmful emissions by the 'best practicable means' — what was 'harmful' being defined in terms of the capacity of the local environment to absorb and, where appropriate, detoxify it.

Each of these approaches had its merits. The one adopted in Britain, it was argued, took account of the technology and investment available at each individual factory and avoided the risk of setting standards that could not be attained and, therefore, would be ignored. The American approach takes account of the quality of the local environment and aims directly to maintain a desired standard. By setting general limits, the European approach treats all industries equally.

The growth in world trade, however, exposed the inadequacies of all attempts to regulate pollution locally. The most obvious danger is that of creating 'pollution havens'. A government which seeks to protect its environment and citizens by enforcing strict emission controls may make its manufactures more expensive than those of another country where emission controls are lax. This will penalize the more environmentally sensitive country in export markets and, unless it raises protectionist barriers, may flood its domestic market with cheaper imported goods. As national companies have merged to form multinational corporations, a further dimension has been added to this danger. The corporations may be tempted to locate their factories in countries where environmental controls are most easily and cheaply satisfied, so that a country seeking overseas investment has a strong incentive to take a relaxed view of the environmental hazards the factories may introduce.

Further trading problems arise when environmental standards differ from one country to another. Motor cars, for example, must be designed to meet certain criteria, but how should a manufacturing company respond when it wishes to produce a car for sale in its home market and for export to, say, 12 countries and faces 13 quite different sets of standards? This might mean, in effect, 13 different models. It would be simpler and cheaper if the criteria were to be standardized so that one model would be acceptable in all of its markets.

Transfrontier pollution also necessitates supranational measures. Scandinavian politicians, for example, complained that the sulphur dioxide which caused the acidification of their lakes was transported across the North Sea from Britain after the British had reduced their own pollution levels by building taller smokestacks. Environmental quality improved in Britain because, by the time the polluting crystals were washed to the ground, they had departed from British shores. Research

revealed this to be a considerable oversimplification of what actually occurred, but the point is valid — and not only for Britain. It may be very difficult for a government to justify to its own electorate the imposition of restrictions on behaviour that produce no discernible benefit for them. The transfrontier problem is, clearly, of great importance in Europe, but an almost identical version of it also arises among the coterminous United States.

United Nations involvement
The solution to such dilemmas is simple in theory but far from simple to achieve. Governments can meet, agree on measures that address clearly defined environmental issues and translate their agreement into appropriate national legislation. The appreciation that this was the most sensible way to proceed grew rapidly during the 1960s, in parallel with the rapid upsurge of popular concern about environmental issues in general and environmental pollution in particular. National governments became increasingly aware of the popular demand for better environmental protection and, by the end of the decade, the United Nations was involved. In June, 1972, the United Nations Conference on the Human Environment was held in Stockholm. One result of the Conference was the establishment of the United Nations Environment Programme (UNEP), a new agency, with headquarters in Nairobi, charged with the task of monitoring the global environment and initiating and co-ordinating international measures for environmental protection and improvement. Since its formation, UNEP has established a global monitoring network to collect and interpret data, and has sponsored many international agreements. The Global Environmental Monitoring System (GEMS) monitors climatic change, the oceans, resource use, human health and the long-range transport of pollutants. Since 1985, the data collected by GEMS have been integrated with information from other UN agencies and then analysed to compile the Global Resource Information Database (GRID). GRID is based in Geneva and uses computers and software developed by the National Aeronautics and Space Administration (NASA). Satellite observations now make an essential contribution to environmental monitoring.

The international study of the atmosphere is co-ordinated by other research programmes. The Global Atmospheric Research Programme (GARP) uses information from meteorological stations, balloon soundings and satellite observations to improve the scientific understanding of the structure of the atmosphere. The Global Horizontal Sounding Technique (GHOST), part of the World Weather Watch, uses polar-orbiting satellites to track balloons that are designed to float at levels where the air density is constant and to collect data from them and transmit them to surface stations.

The improvement of air quality has become a global issue, in the sense that local or regional action must be related to a much wider context. The role of UNEP, therefore, is as much political as it is scientific. It is not only information the agency must co-ordinate, but the responses of governments to it. The discussions which precede international agreements usually have two components, both of which take place under UNEP auspices. Scientific advisers to their governments meet in working groups to analyse the issue they are to address. This may require them to organize and then supervise research projects. They agree on steps that are scientifically desirable and feasible and report to their governments. Then the politicians meet to translate the scientific recommendations into international agreements in a series of steps. A draft treaty is prepared. When this has been agreed and signed it becomes a protocol (the Greek *protos* means 'first' and *kolla* means 'glue'). The signatories later confirm — ratify — their agreement, after which the matter is concluded and the protocol becomes a treaty.

The issues that are discussed are seldom uncontroversial. It may appear obvious that particular substances are harmful to human health or to the environment, but such claims are often open to challenge, and politicians must reach their decisions by balancing conflicting or sparse scientific evidence.

The Montreal Protocol on Substances that Deplete the Ozone Layer, drawn up in 1987, is the most publicized of the international agreements UNEP has sponsored, although it is not the only one. The Montreal Protocol commits governments to legislate to reduce the production and use of CFCs by agreed amounts by agreed dates —

and the phasing out of CFCs has been accelerated since the Protocol was signed. Work on the Protocol has been followed by the much more difficult task of finding an equally comprehensive agreement on ways to minimize the risk of climate change induced by human activities.

The issues dealt with by such agreements are only superficially simple. The scientific basis for them is seldom firm. When the Montreal Protocol was being prepared, no one could say with confidence that a depletion of the ozone layer was more than possible, or that CFCs would certainly trigger depletion, or even that a thinner ozone layer would be biologically harmful. The scientists needed more information but warned that, by the time they obtained confirmation (or denial) of their fears, it might be too late for effective remedial action. Politicians decided it would be irresponsible to take the risk and, therefore, they had to face the political complexities. CFCs were used as aerosol propellants (they have now been replaced in this use almost entirely), but their main use is as refrigerants and foaming agents, and related halogen compounds are used in fire extinguishers. The compounds are completely non-toxic and non-flammable and the industrializing countries were planning to increase greatly their use of them, mainly for refrigeration. In tropical climates, where food spoils quickly, refrigerators can make an important contribution to standards of living and health. Alternatives were not immediately available and promised to be more costly, so countries such as India and China refused to accept the Protocol unless they were promised economic and technological help that would allow them to make the transition.

Formidable obstacles were overcome to secure agreement to phase out CFCs, which comprise a limited range of specialized products. Much greater difficulties face those who would reduce the risk of induced climate change and, quite apart from the political and economic aspect of the negotiations, there are many scientific uncertainties that will be resolved only after years of painstaking research. There is no disagreement about the principle involved — certain gases absorb long-wave radiation and, therefore, changes in their atmospheric concentrations have climatic implications — but there is great uncertainty about the extent to which climates may change and the local effects such change may bring. Without this information, we risk asking politicians to find answers to questions that have not been adequately formulated.

The risk of climatic change differs from all other environmental issues. Obviously, it differs in scale. Many other issues have world-wide implications, but no other aspect of our interference in the natural systems that govern the planet threatens such fundamental change. It also differs in that the release of the major greenhouse gases is a direct consequence of our entire way of life. Carbon dioxide is emitted when we burn fuels for energy, for example, and methane is a by-product not simply of European and North American farming but of the cultivation of rice in Asia — and in terms of the number of people relying on it, rice is the world's most important staple food.

Consequences of a rise in temperature

It is predicted that, by the second half of the next century, the average global air temperature may have increased by about 8 °F (4.5 °C) and that, unless the atmospheric accumulation of greenhouse gases has been checked, it will still be rising. The increase is predicted to affect mainly the higher latitudes, leaving the Equatorial regions little altered. Warmer weather frightens no one, but this is only the first predicted effect and the cause of many others. The governments of many low-lying Pacific islands, for example, fear that a rise in sea level will inundate their lands. It seems self-evident that, as the oceans warm, their waters will expand and that their volume will increase as polar and glacial ice melts. Sea levels are rising now, at about 1 inch (25.4 mm) a year, due partly to isostatic readjustment and partly to expansion due to warming and, were all the ice caps to melt completely, the sea might be 260 feet (80 m) higher than it is today, reducing Britain to a group of islands. This is not a genuine risk but neither is a change of this magnitude fanciful. During the most extreme period of the Pleistocene ice ages, the sea level around Britain was 425 feet (130 m) lower than it is now.

A rise in temperature will lead to increased evaporation and, therefore, increased precipitation — including precipitation in polar regions. This suggests that the polar ice caps will grow thicker, not thinner. Although the temperature at the surface of polar ice sheets may increase, their high albedo will ensure the warming does not penetrate. Ice and snow which melt at the surface will flow down small cracks and crevices and freeze again before a significant amount can reach the sea. The sea-level rise is likely to be much smaller than was once feared — the current prediction for the second half of the next century is about 26 inches (650 mm) and the rise may not be spread uniformly. Much more extensive melting of the ice caps, if it occurs, will be delayed until long after the end of the twenty-first century.

Even so apparently small a rise may have important consequences, however. The Pacific islands are not the only populated regions of the world that lie close to the present sea level. So do many of our major cities — built originally as sea ports — and so do large land areas, including most of the Netherlands and much of the eastern seaboard of the United States. Other coastal regions, such as parts of the east coast of England, may suffer increased erosion. Vulnerable coasts can be protected, although the cost of building the necessary structures would be high. South-east England is sinking (and Scotland and Scandinavia are rising) as part of the adjustment of land levels — 'isostatic readjustment' — following the melting of the ice at the end of the most recent ice age. The Thames Barrier was constructed to protect low-lying London from the risk of floods due to the combined effect of falling land level and tidal surges, which are caused by high tides following several days of strong onshore winds. Its planners did not allow for a sea-level rise due to the greenhouse effect, but the Barrier could be raised. There is a more insidious risk, however, of salt water contaminating fresh groundwater and rendering infertile what is now rich farming land. In the Netherlands, selected areas have been deliberately flooded with fresh water to raise the water table above the sea level, and so prevent the incursion of salt water below ground level.

The locations of the broad vegetational regions of the world are determined by regional climates. If the global climate were to change, those locations would be shifted. Essentially, the climates and vegetation associated with the subtropics might be

The Thames Flood Barrier can be closed to protect the city from the combined effects of a high tide and storm surge caused by several days of strong onshore winds in the Thames Estuary.

expected to expand polewards, compressing all of the belts in higher latitudes. This kind of change has happened many times before, during the glacial advances and retreats of the Pleistocene, and living organisms have responded by migrating from old to new areas. The process is well enough documented for scientists to have calculated the rate at which plants migrated, with the more mobile animals following them. No doubt this would happen again, except that today humans occupy and use such a large proportion of the total land in the temperate latitudes as to constitute a barrier to free migration. If the extinction of many species is to be averted, it may be necessary to provide areas and meridional corridors of undisturbed reserves to permit migration.

The total area of tundra would be much reduced because it lies at the extreme poleward limit and can be compressed no further. The loss of much of the tundra might seem to be of little importance, but its replacement by conifer (boreal) forest would involve the melting of vast areas of perma-frost — permanently frozen ground. This, it is feared, would provide a large biochemical

stimulus leading to the release of methane by bacteria and so exacerbating the green-house problem. Other scientists have conducted experiments, however, that lead them to suggest the methane would be utilized by other bacteria as rapidly as it was produced. It is one of the areas of uncer-tainty — but very important.

The familiar factory chimney was invented in the last century, by James Muspratt (1793-1886), as a device to reduce air pollution. The pollutant that concerned him was hydrochloric acid, a waste product from the manufacture of 'alkali' — sodium carbonate, which is also known as 'soda ash' and 'washing soda' — a necessary ingredient for the making of soap. The acid killed vegetation around the factory but the chimneys, almost 300 feet (91.5 m) tall, dispersed it over a much wider area and, therefore, diluted it. The dilution helped, but did not solve the problem, because the acid continued to be discharged into the air. A little later, a process was invented that collected the acid in liquid form. It was no longer released into the air but there was no demand for hydrochloric acid so it was dumped at sea, until the invention of

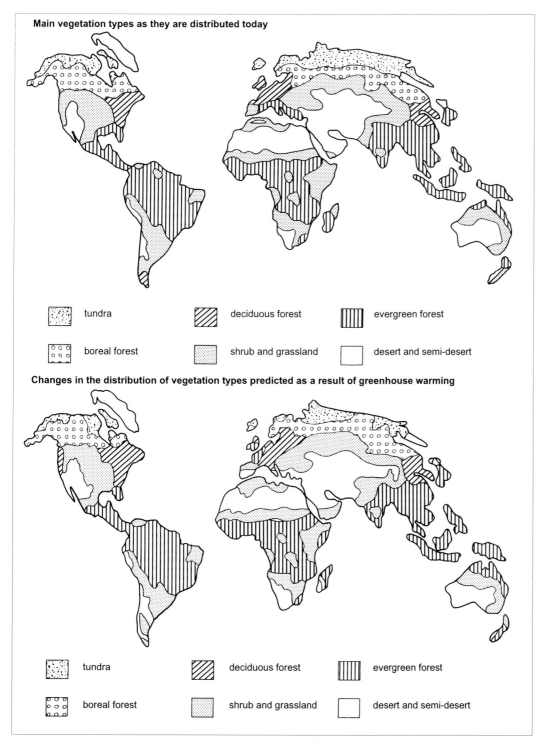

Main vegetation types as they are distributed today

tundra   deciduous forest   evergreen forest

boreal forest   shrub and grassland   desert and semi-desert

Changes in the distribution of vegetation types predicted as a result of greenhouse warming

tundra   deciduous forest   evergreen forest

boreal forest   shrub and grassland   desert and semi-desert

Vegetation and the greenhouse effect. A global warming may expand the region occupied by vegetation that is now typical of the middle latitudes, largely at the expense of high-latitude tundra. Cereal growing will become possible further north in America and Asia. Some deserts may be larger, and desert may extend into southern Europe, but the Sahara and Arabian deserts will be smaller and the total area of cultivable land may increase in the world as a whole.

further processes led to the separation of chlorine from the acid and its reaction with calcium hydroxide to produce bleaching powder, a new product which could be marketed profitably.

This story is far from exceptional. In the past, many pollution problems have been solved in this way. Where a pollutant is the unwanted by-product of an industrial process, sooner or later an imaginative chemist or engineer is quite likely to devise some way of converting it into a marketable product. There is a limit, however, to the use of the by-products of one process as raw materials for another. The waste may have no conceivable use, for example, or it, together with everything that can be made from it, may already be commonplace and inexpensive, so no market exists or can be created and the older the industrial process that yields the by-product, the more likely it is that this will be so.

Nevertheless, industrial history contains so many accounts of companies that have turned potential failure into phenomenal success by down-grading their traditional products and using what were formerly wastes to make exciting new products, that the dream is deeply ingrained. No corporation is so large that its directors would not feel a frisson of sheer delight at the prospect of climbing yet higher, as it were, by pulling themselves up by their own bootstraps.

If a commercial use can be found for a polluting waste, we may be reasonably confident that it will be exploited. The residual 'by-product' pollutants are those for which there is no use, and their emission can be controlled only by requiring them to be captured and consigned to a safe form of disposal.

Sulphur dioxide, for example, is now regarded as a serious atmospheric pollutant — although sulphur is an essential nutrient for plants and, while acid precipitation causes harm in some places, in others plants benefit from it. It can be washed out of the waste gas to produce sulphuric acid, but there is no shortage of sulphuric acid. The acid cannot be sold and disposing of it safely is expensive. The more usual method removes sulphur dioxide by passing a waste gas containing it through a bath filled with a suspension of calcium hydroxide (slaked lime) in water. The sulphur dioxide reacts with the calcium hydroxide to yield calcium sulphate, which is insoluble and can be collected. Calcium sulphate, which is chemically identical to the mineral gypsum, is environmentally harmless and can be dumped in landfill sites — gypsum has many commercial uses but the market for it is amply supplied.

It often happens, however, that in solving one problem we create others. The removal of sulphur dioxide is not difficult, provided there is a sufficient supply of calcium hydroxide. Calcium hydroxide [$Ca(OH)_2$] is made by adding water (H_2O) to calcium oxide (CaO) — 'slaking' it. The oxide is obtained by heating ('kilning') calcium carbonate ($CaCO_3$) to drive off the carbon dioxide (CO_2) — but carbon dioxide is a greenhouse gas and, therefore, removing sulphur dioxide contributes to the greenhouse effect. Calcium carbonate is obtained in the form of quarried limestone rock — but many of the most attractive landscapes are found in limestone areas.

The remedies and what they may imply

We might reduce emissions of sulphur dioxide still more simply were we to burn less of the fuels that contain it. Even so attractive a proposition should be approached with a little caution, however, for it is not entirely free from environmental cost. Oxidized to sulphate, sulphur dioxide encourages clouds to form, thereby shading the surface and helping to offset greenhouse warming. If we burn less fuel we will emit less of both sulphur dioxide and carbon dioxide. The reduction will tend to exacerbate warming, but also to alleviate it. Which effect dominates depends on the average length of time a molecule of each substance resides in the atmosphere. Sulphur dioxide remains for only a matter of days, but carbon dioxide remains for many years and, therefore, the sulphur dioxide concentration will fall faster than that of carbon dioxide. If we reduce our use of carbon-based fuels containing sulphur, there will be a lag of some years during which we aggravate the greenhouse effect before we begin to diminish it. This is not an argument for abandoning the idea of burning less carbon fuel, but merely an illustration of the way in which apparently obvious remedies to problems can have surprising implications.

Not all industrial wastes can find a use, but there is a general sense in which pollution often amounts to waste and might be remedied by greater efficiency. This applies even to substances that enter the air as a consequence of their ordinary use. Pesticides, for example, are commonly applied as sprays, but they can be effective only if they reach their intended targets. If they drift away in the air to cause pollution they are being wasted. More efficient application — and appropriate equipment now exists — uses less of the chemical with much greater precision. Spraying becomes cheaper and pollution is reduced at the same time. Once in the soil, nitrogen-based fertilizers can engage in chemical reactions that release gaseous oxides of nitrogen and ammonia. More efficient fertilizer use would reduce pollution and also save money for the farmer.

Thrift may provide a more compelling reason for reducing pollution than altruistic appeals on behalf of the welfare of the planet or of human generations yet unborn. The greatest source of atmospheric pollution is the burning of fuel and, no matter how ingeniously we devise ways to burn it more cleanly, if we can learn to burn less of it we will profit economically and immediately as well as environmentally, but in the long term. In some cases there may also be social benefits.

Were we to rely more on public transport, for example, and less on private cars, there might be a significant saving in the amount of fuel used for transportation. This would produce environmental benefits, not only in terms of pollution — especially if more vehicles were powered by electric motors — but also by reducing traffic congestion in cities. More space on the streets would make it easier to improve facilities for pedestrians and cyclists, neither of which causes any pollution at all. At the same time, better public transport systems would provide mobility for the large number of people who have no ready access to

This low-energy house, in Milton Keynes, England, uses an array of photovoltaic cells to generate some of its electricity.

A fuel cell uses two 'fuels', commonly hydrogen and oxygen, separated by porous plates. A charge applied to the plates causes the fuels to diffuse through the plates. The oxygen plate acquires a positive charge, the hydrogen plate a negative charge, and so electrons flow around the circuit.

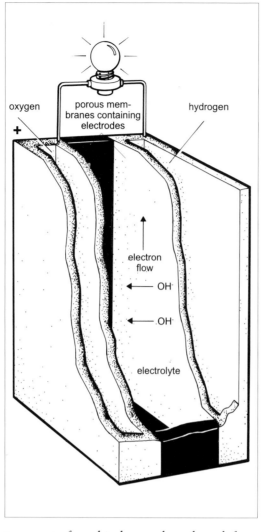

oxygen

porous membranes containing electrodes

hydrogen

+

electron flow

OH⁻

.OH⁻

electrolyte

of condensation and evaporation in a working fluid that is liquid when compressed and gaseous when the pressure on it is relieved. The fluid vaporizes in a cold area, such as the exterior of a building, extracting the latent heat of evaporation from the surrounding air. When compression causes the gas to condense, the latent heat is given up and warms its surroundings.

Apart from using less fuel, we may need to increase our use of alternatives to the carbon-based fuels. A fuel cell may be one alternative. It is silent, has no moving parts and, in principle, it is by far the most efficient way to generate electricity. The idea is not new. The first working fuel cells were constructed in 1839 and, since then, a great deal of research has been devoted to producing a type of cell that can be fitted in vehicles as well as supplying a stationary source of power. Fuel cells have been used in manned spacecraft.

When fuel is burned to generate electricity, the chemical energy of the fuel is converted to heat and the heat operates a dynamo. A fuel cell also converts chemical energy to electrical energy but it omits the intermediate stage, of producing heat. In its simplest form, a fuel cell consists of a box with three chambers. One chamber contains the fuel — usually hydrogen but other fuels can be used — and is separated from the central chamber by a porous membrane that incorporates a cathode (positively charged electrode). The central chamber contains an electrolyte — a liquid that conducts electricity — and it is separated from the third chamber by another porous membrane incorporating an anode (negatively charged electrode). The third chamber contains oxygen or air. The hydrogen reacts at the cathode with the electrolyte, releasing electrons. The oxygen reacts at the anode with the electrolyte, absorbing electrons, the flow of electrons being carried through a system of wiring. A flow of electrons constitutes an electric current.

A fuel cell works efficiently on a small scale, but it is difficult to design a system of cells for large-scale generation. Despite the obvious attractions, fuel cells have disadvantages. They are inclined to produce only weak currents. Their performance can be improved by raising the working temperature — most electrolytes must be heated in any case — and by including a catalyst in

transport of any kind — such as those left at home by the person who commutes to and from work in the family's only car.

We burn a great deal of fuel to heat buildings and to operate air-conditioning plant. It is perfectly feasible to design buildings that require little or no fuel for these purposes and demonstration 'low-energy' houses and schools have existed for many years. They are new buildings, of course, and even if all new buildings were built to such designs and with the appropriate materials — which they are not — it would take at least several generations to replace the entire stock of buildings. Meanwhile, some improvements are feasible within existing buildings. Improved insulation will help and so will more efficient appliances.

The heat pump, for example, is simple and efficient. It consists of a small electric motor which pumps a fluid around a system of pipes, compressing it at one stage in the cycle. The device exploits the latent heat the electrodes, but this increases their cost and the cheaper fuels, such as hydrocarbons, require the more expensive catalysts. Nevertheless, work on them is continuing and there is reason to believe that the problems associated with them can be overcome.

For the time being, however, the prospects remain uncertain for devices such as fuel cells, and the economies of which we are realistically capable may be insufficient to achieve a substantial reduction in our use of carbon fuels. We can make some use of 'biomass' fuels — produced from plants grown for the purpose. They release only the carbon dioxide that was absorbed by the growth of the plants, but they occupy land that might otherwise produce food and, in much of the world, farm land is a scarce resource. In the end, if we are persuaded that the protection of the atmosphere necessitates a major reduction in our emissions of carbon dioxide, we will have no alternative but to accept the nuclear generation of electricity and increase our reliance on it.

The bicycle is said to be the most efficient vehicle ever invented — and the cleanest. Its popularity is well deserved, and continues to grow.

PUTTING AIR TO WORK

Animal flight

Should you ever travel by sea in the tropics or subtropics, you may be fortunate enough to encounter flying fishes. You might even see an entire shoal of them leap from the ocean and glide effortlessly, skimming the wave tops, at about 10 miles per hour (16 km/h) for 200 yards (183 m) or more before returning to the water. If they were members of the commonest species, *Exocoetus volitans*, or if they were *Cypselurus heterurus* (the Atlantic flying fish) you might recognize them by their large pectoral fins, which they spread like wings as they take to the air. A flying fish will often prolong its glide by dipping its tail in the water and beating it vigorously to regain flying speed.

Probably, the shoal was being chased by predators, which drove the flying fishes to swim faster and faster until their speed was sufficient for them to elude their pursuers by taking off and entering another medium. It is a very effective means of escape and more than 100 species of flying fishes have adopted it. So have the hatchetfishes which live in South American rivers (they belong to the family Gasteropelecidae and are not related to the deep-sea hatchetfishes of the family Sternoptychidae). Hatchetfishes flap their fins as well as their tails and, therefore, come close to powered flight, although it does not carry them very far. One member of the family, the keeled belly (*Gasteropelecus sternicla*), is a popular aquarium fish — and the aquarium should always have a lid!

It is not only fishes that leave the water in this way. There is a family (Onycoteuthidae) of flying squids that can propel themselves out of the water like missiles, to a height of 12 feet (3.7 m) or more, and members of one species, *Dosidicus gigas*, glides at up to 16 miles per hour (25.7 km/h).

The ability to take to the air may also have other uses, however. Some of the halfbeaks also fly — although they do not have enlarged pectoral fins. Relatives of the flying fishes, their name is derived from the fact that the lower jaw is much longer — in some species three times longer — than the upper jaw, and the halfbeaks are believed to use it to scoop up small fish as they skim along, just above the surface. For them, what may have begun as a means of escape has become a hunting technique.

Strictly speaking, flying fishes do not fly, they glide. Nevertheless, true flight, in which the use of muscular power permits an animal to remain airborne for much longer, is a popular evolutionary development. It has appeared independently — which means starting afresh each time — in insects, reptiles, birds and possibly twice in mammals. If we use a wider definition, however, that embraces any living organism which routinely flies, glides or floats through the air, the list of aviators is very long indeed.

The list is not confined to animals. As hay fever sufferers are only too well aware, many plants, including the grasses of which there are about 10,000 species, rely on the wind to distribute their pollen. Stand outdoors in summer and the air may be filled with thistledown — plant seeds that disperse by drifting wherever the breeze may carry them. By dispersing pollen or seed, the air plays an essential role in their reproduction. Were we to widen the definition even further, to include passengers, many more plants would qualify. We would have to count all those which are pollinated by flying insects and those which rely on birds to distribute their seeds.

All these are flowering plants, a group (Angiospermae) that evolved only recently — about 130 million years ago — but they were by no means the first organisms to exploit the possibilities of air travel. Tap the ripe fruiting bodies of many fungi — fungi are no longer classified as plants, but occupy a taxonomic kingdom of their own, the Fungi — and they will release what looks like a cloud of fine dust. Each 'dust' particle is a spore, capable of developing into a new fungus. Indeed, it may well have been by air that the first organisms travelled from the water to colonize dry land. Bacteria, among the most ancient of all organisms, also produce spores and they may have been the first land arrivals.

The fishes and squids have existed for a very long time and we cannot know when it was that they first took to leaping from the water and gliding. What is more certain is

(Above) As the view from space shows, the Earth's albedo is determined very largely by the amount of cloud cover. Should the climate grow warmer, more water will evaporate and more cloud will form. The extent to which increased cloud will contribute to warming (because water vapour and droplets absorb long-wave radiation) or offset it (because of the increase in albedo) is the subject of much scientific debate. (Left) Many species of lichens (but not all) are very intolerant of sulphur. If you find them growing abundantly, you can be certain the air is not seriously contaminated by sulphur dioxide.

(Right) The Equatorial regions support the most luxuriant vegetation on Earth. The intensity of UV radiation at the Equator is about seven times that in temperate latitudes, but it seems not to inhibit plant growth. (Below) Much of East Anglia, England, is low-lying and could be inundated by a rise in sea level.

that the earliest airborne land animals were probably spiders, of which the oldest fossils date from about 380 million years ago. They quickly adapted to life on land, although a few returned to the water later. All spiders produce silk — although not all of them use it to build webs. Most species of spiders enclose their eggs in silken sacs, which is probably the original use for their silk, and guard their young, which hatch as 'spiderlings' — miniatures of their parents. A single female may produce many young and, before long, the spiderlings must disperse in search of territories in which to hunt prey. They often do so by 'ballooning'. A tiny spider climbs as high as it can on a plant, then spins a silken thread that drifts with the wind until it is long enough to lift the spider bodily and carry it high into the air. Ballooning spiders sometimes reach altitudes of more than 3000 feet (915 m) and may drift for more than 60 miles (97 km). The commonest spiders in the northern hemisphere are the very small 'money spiders' (family Linyphiidae) and most of them migrate in this way.

Falling cannot injure them. Animals as small as spiders or insects have so low a mass that gravity hardly affects them — they are barely heavy enough to fall at all and are assured of a gentle landing regardless of the height from which they descend.

Insect evolution

Insects have taken this exploitation of the air much further by acquiring wings. They appeared on Earth at about the same time as spiders, and the oldest fossils include both winged and wingless forms but no intermediate forms, with partly evolved wings, and so no one can be sure just how insects acquired their wings in the first place. The most likely explanation is that certain very early jumping insects had flanges projecting to the sides of the upper parts of some of their body segments. These flanges would have ensured that its jumps always ended with the insect the right way up. Over the course of many generations, the flanges became larger, until the insects could glide and, finally, they ceased to be smooth continuations of the exoskeleton — the exterior 'skin' of an insect which gives its body rigidity — and became attached to it by hinges. Once the wings were capable of movement, insects were able to fly.

This was a major development and,

evolutionarily, it was immensely successful. In terms of individuals and of species, insects outnumber all the other animals on Earth put together, and by a huge margin. Even by the most conservative estimates there are at least 750,000 species of insects, including 350,000 species of beetles, 165,000 species of butterflies and moths and 85,000 species of flies. In the course of their evolution, some species have lost the wings their ancestors possessed — they are 'secondarily wingless' — but there are less than 600 species that are 'primitively wingless' — meaning they never developed wings.

Insect evolution did not end with their acquisition of wings, of course. New groups of insects continued to appear bearing improved and strengthened wings. An insect wing grows from the exoskeleton and is made from two layers of 'skin'. The veins, so characteristic of an insect wing that in many species their patterns are an important feature in their identification, are made from thickened 'skin' enclosing a tube that connects to the interior of the body and through which blood flows. In evolutionarily more primitive insects, the veins form a network, but in more advanced forms they are reduced to a small number running along and across the wing, rather like struts, thereby increasing its strength. At first, the wings were held outstretched when the insect was at rest, as they are still by dragonflies. A further evolutionary development allowed later insects to fold their wings back over the body when at rest. In this position the wings are out of the way while the insect is resting or feeding and they are better protected against damage.

The wing is attached to the upper surface of the body segment — the 'tergum' — from which it grows and the edge of the side of the segment — the 'pleuron' — acts as a fulcrum, pressing against the wing a little way from the point of attachment. When vertical muscles — known as 'indirect flight muscles' — contract, the tergum is pulled downwards and, therefore, the wing moves upwards. In some insects, relaxing the muscles and raising the tergum causes the wing to move downwards, but in others downward movement is produced by muscles attached to the wing itself — called 'direct flight muscles' — and there are also insects which use a combination of the two methods. The wings move back and forth as well as up and down, turning as they do so to alter the angle with which they meet the air to provide lift, acting upwards, and thrust, acting forwards, at the same time. The result, varying from one insect group to another, is either an elliptical or a figure-of-eight motion.

When insects first evolved wings, they evolved two pairs of them and, because of the way the wings operate, each pair was originally capable of moving independently of the other. Many insects retain both pairs but there was a drawback to independent movement. The movement of any object through the air causes turbulence and the hind wings had to beat in the turbulent air produced by the fore wings. Some insects, such as damselflies, continue to fly in this way, but the result is a loss of efficiency in the hind wings. Damselflies are rather weak fliers. Dragonflies, their close relatives, evolved a system in which the movement of the two pairs of wings is co-ordinated. As the fore wings move upwards, the hind wings move downwards, avoiding the worst of the turbulence. Other groups of insects have overlapping fore and hind wings or, as in butterflies and moths, the two sets of wings are coupled together by hooks. These modifications allow the four wings to function as a single pair.

In some insects, one pair of wings has been modified for other uses, so resolving the difficulty. Beetles use only their hind wings for flying. The fore wings have become 'elytra', the hard cases that fold over and protect the wings when the insect is not using them. The 'true' flies — a group that includes the gnats and mosquitoes as well as houseflies and bluebottles — use only their fore wings to fly. Their hind wings have become 'halteres', small knobs that beat with the same frequency as the wings and help to stabilize the flight.

Flying ability varies widely but, at its best, it is very impressive. Consider, for example, the complexity of the movements performed by an ordinary housefly as it turns to land upside down on a ceiling, its wings making about 200 complete beats a second, or a cloud of midges, hovering without ever colliding, each insect beating its wings at up to 1000 times a second. Some insects are also capable of considerable speeds over short distances. Horseflies and some dragonflies, for example, can fly at more than 30 miles per hour (48 km/h). Others have

Using its halteres as stabilizers, a hoverfly is able to to achieve hovering — a most difficult feat. Its banding is a disguise. The hoverfly is a true fly, with one pair of wings, not a wasp which has two pairs.

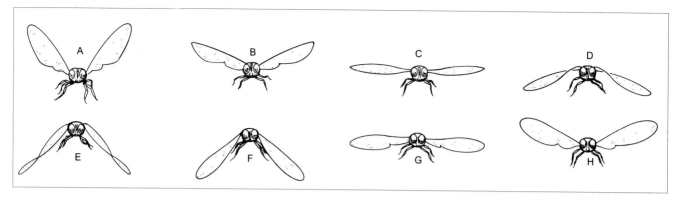

great endurance. Honeybees will travel up to 30 miles (48 km) and, during their seasonal migrations, monarch butterflies travel more than 1500 miles (2413 km), covering up to 650 miles (1046 km) without needing to feed.

Flight is so commonplace and has evolved so frequently that it is reasonable to suppose we might find flying animals on any Earth-like planet that supports life. So we might, but only if the planet has an atmosphere closely resembling that of Earth. Flight on Earth is possible only because 21 per cent of our atmosphere is oxygen. The oxidation of carbohydrates provides the large amount of energy that is needed to sustain an animal in powered flight — and explains why the diet of so many flying animals is rich in sugars. The 'engine' of a desert locust, for example, produces about 57 joules of energy for each gram of its body weight (386 calories per ounce) per hour. If the air contained much less oxygen, food could not be oxidized rapidly enough to supply the energy for flight. One day it may be possible to establish some forms of life outdoors on Mars but, unless the martian atmosphere is supplied with a great deal more oxygen than seems possible, no bird will ever cross the martian skies — and there will be no pollinating insects.

The chemical reactions that supply the energy require that the oxygen be transported by the blood, and it is the means by which oxygen enters the blood that has imposed an upward limit on the size of insects. The largest insect alive today is probably the Goliath beetle (*Goliathus giganteus*) which is about 6 inches (150 mm) long, but the largest insects ever known lived some 300 million years ago, near the end of the Carboniferous. They were dragonflies and one of them, *Meganeura gracilipes*, had a wingspan of more than 28 inches (700 mm). This is about the

greatest size an insect can attain.

Science fiction tales of insects the size of humans are just that — fiction — because such an insect would asphyxiate. Although many aquatic insect larvae possess gills, oxygen passes into the body of an adult insect through small tubes — 'tracheae' — that open through pores in the exoskeleton and then diffuses directly into the blood. The blood fills most of the body cavity and the 'heart' serves merely to keep it well stirred. Diffusion is very efficient but only over short distances. Over longer distances it is too slow and all larger animals have evolved gills or lungs, which provide a very large surface across which oxygen can enter the blood, and a more complicated and powerful heart to pump the blood around a closed system of blood vessels.

Flying reptiles

This is the method used by all vertebrate animals, and the first of the vertebrates to fly were the reptiles — and they did so twice. They first took to the air about 200 million years ago, with the appearance of the pterosaurs, and the flying reptiles reached their evolutionary peak around 50 million years later. By the end of the Cretaceous, 65 million years ago, all of them had become extinct but a second reptilian group had evolved into the birds.

Most pterosaurs were not especially large. *Rhamphorhynchus*, a typical representative, was about 18 inches (457 mm) long, including a very long tail ending in a small rudder. It had large eyes and long jaws with teeth that sloped forwards, presumably for spearing fish. Many of the bones were hollow and filled with air, like those of birds, to reduce the body weight and the wings were carried on the fore limbs — the arms.

Like all land-dwelling vertebrates, pterosaurs had four limbs, each of which ended in a 'hand' or 'foot' with five digits,

The wingbeat cycle of a fly. The wings describe an approximately circular motion, pushing forwards and downwards, then backwards and upwards, and are twisted as they move so the angle between the wing surface and the airflow (the angle of attack) changes constantly. The overall effect is to produce a wash of air downwards and to the rear so that the wings provide a forward thrust as well as upward lift.

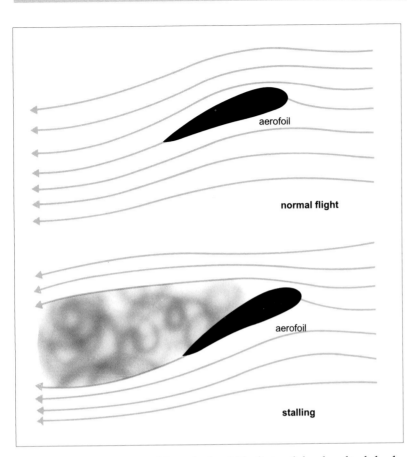

aerofoil

normal flight

aerofoil

stalling

Airflow and pressure distribution over an aerofoil. When air moves smoothly across an aerofoil (top), the pressure is lower on the upper surface than on the lower. This produces an upward force, lift. If the angle between the aerofoil and the air flow (the angle of attack) is increased beyond a critical point (bottom), the smooth flow of air breaks down, lift is lost and the aerofoil stalls.

(Opposite) When climbing rapidly, pigeons are able to twist their wings to reduce resistance and provide some lift on the upstroke, as well as drag, but because the wings are moving to the rear, the drag is directed forwards, as thrust. At the top of the upstroke they may clap their wings together and then open them rapidly. This also generates lift.

although the fifth digit of the fore limb had disappeared in the course of evolution. Three of the 'fingers' were typically reptilian and had claws, but the fourth was immensely long and provided the whole of the support for the wing, the inner side of which was attached to the body all the way from the neck to the thigh. The articulation of the hind legs made them point to the rear and the animal could not have walked with them. When it was not flying, presumably it hung from a tree or rock ledge by its fingers or by its feet, upside down like a bat.

The wings were made from skin and the animal is believed to have been covered in hair. One pterosaur fossil, *Sordes pilosus*, that was found in the USSR in 1971, was covered in thick fur. The fur would have provided insulation and the pterosaurs may have been endothermic — 'warm-blooded'.

Pterosaurs certainly flew, but it is difficult to see how the structure of their wings would have allowed them to do more than flap them up and down — and they were very vulnerable to damage. It is more likely that they soared, diving to gain speed and seeking updraughts to gain height. There were many species of pterosaurs and a second group of flying reptiles, the pterodactyls, evolved from them. They were broadly

similar, but with much smaller tails, and ranged greatly in size. Some were no bigger than a sparrow but *Pteranodon*, with a wingspan of up to 25 feet (7.6 m) was the largest animal ever to fly under its own power — its wingspan was double that of today's wandering albatross.

Birds

Apart from the pterosaurs, most reptiles were covered with scales and, in one group, the scales evolved into feathers. Feathers are the only part of its anatomy that unambiguously distinguishes a bird from its archosaur — 'ruling reptile' — ancestors. *Archaeopteryx lithographica*, which lived about 140 million years ago, had the teeth, long tail and claws of a reptile together with such bird-like features as feathers, a foot adapted for perching and a fore limb that was fully modified into a wing, using all the digits for support rather than just one. *Archaeopteryx* flew, but there is some doubt about how strongly it was able to do so, because it appears to have lacked the strong muscles that operate the wings of a bird.

Feathers are believed to have evolved as thermal insulation to help maintain a constant body temperature in endothermic reptiles. Their aerodynamic properties became evident later, when the 'warm-blooded', feathered reptiles took up aviation. They are strong, flexible, waterproof, windproof, weigh little — they are 'light as feathers' — and they are easily repaired or replaced. A small tear amounts to serious damage in a wing that consists of a membrane, but a damaged feather or two is of little consequence. Aerodynamically, however, feathers contribute more, because they provide a smooth, streamlined surface that can be shaped to make an aerofoil.

Aerodynamics

It was Daniel Bernoulli (1700-82), the Swiss botanist, anatomist and mathematician, who discovered that when a fluid flows through a pipe that has a constriction, the constriction causes the fluid to accelerate as it passes and the pressure in the fluid to decrease. Expressed mathematically, this is known as 'Bernoulli's theorem' and, as well as its many applications in hydrodynamics, it forms the basis of the science of aerodynamics. Think of the way air flows past an object that is rectangular in cross-section — like a plank of timber. The air encounters a

The wings of a bat consist of a membrane of skin joined to the body, tail and hind limbs, and supported by the fore limbs and their greatly elongated fingers. This is a greater horseshoe bat (*Rhinolophus ferrumequinum*).

surface at right angles to its direction of flow. Some of the air is deflected upwards, some downwards and some backwards and the flow becomes turbulent. Now smooth the plank so the edge meeting the air — the 'leading edge' — is rounded. The air flows much more smoothly. Make the upper and lower surfaces of the object rounded and the flow will be smoother still, but the air will be accelerated as it passes the obstruction and, as it accelerates, the pressure diminishes. Finally, make the upper surface much more rounded than the lower surface. Then the air will be accelerated more over the upper surface than over the lower surface and, therefore, the pressure over the upper surface will be less than the pressure on the lower surface. This will produce a force acting upwards on the whole object. The force is called 'lift', and an object with this cross-sectional shape is called an 'aerofoil'. The wing of a bird is an aerofoil of a very sophisticated kind that allows birds to glide and soar with little effort.

Other aviators

Birds have colonized every conceivable type of habitat. Many have learned to float on the water like boats, dabbling or diving

beneath the surface for food and some, such as cormorants, have learned to swim. Others have taken to living on the ground.

Birds are not the only vertebrate aviators, of course. They share their aerial domain with the bats. No one knows precisely when bats first appeared because, at first, they differed little from their ancestors, which were small, tree-dwelling, insectivorous mammals. Certainly they had taken to the air by about 54 million years ago. Today, there are about 700 species of small bats, the suborder Microchiroptera, and more than 170 species of 'flying foxes' or 'fruit bats', the Megachiroptera. These appeared later, about 30 million years ago, and zoologists now believe they are related more closely to the primates and flying lemurs (Dermoptera) than to the Microchiroptera. If this is so, then flight has evolved independently in two groups of mammals. The wing of a bat is made from a skin membrane supported by all the digits of the fore limb, and it allows the animal great manoeuvrability.

There are many more animals — the gliders — which have progressed some way in the direction of flight. Watch a squirrel leaping from branch to branch and it seems that with only a small improvement it would be able to glide much further, from tree to tree, rarely needing to visit the ground. The flying squirrels of Africa and Asia can do this, by means of a flap of skin between their fore and hind limbs which spreads out when they extend their legs. The flying phalangers of Australia and part of Indonesia use a similar technique. They are marsupials and have evolved by a separate route from the placental mammals. The colugos or 'flying lemurs' of Asia have adapted so well to life in the trees that they are almost helpless on the ground — yet they can glide between trees for up to 450 feet (137 m).

Nor are the mammals the only forest dwellers among which parts of the body have evolved to permit gliding. There are also gliding amphibians and reptiles. A family of frogs (Rhacophoridae), found throughout the Old World tropics, has developed greatly enlarged feet, with webbed fingers and toes. Using the webs as gliding membranes, they launch themselves from tree to tree with outspread limbs and, in effect, a wing at the end of each. *Draco volans*, the flying dragon or flying lizard of Asia, has a long tail for balance and can move some of its ribs sideways, like a pair of wings. There is even a genus of tree snakes, *Chrysopelea*, that has taken to gliding. The snake makes the under side of its body slightly concave and undulates as it glides.

Human flight

Daedalus was the most skilled craftsman of his day. Among other things, he designed and built for King Minos of Crete the labyrinth from which Theseus escaped with the help of Ariadne. Kings are not the most reliable of employers, however, and eventually Daedalus fell from favour and was imprisoned in a tower. Escaping from the tower was not difficult for one of his talent, but Crete is an island and Minos had posted guards to monitor all maritime traffic. Unable to leave by sea, Daedalus determined to fly and set about making two pairs of wings, one for himself and the other for his small son, Icarus. The wings were made from feathers. Daedalus began with small feathers, which he fastened together with wax, then stitched progressively larger ones

The flying squirrel (this is *Glaucomys volans*) glides from tree to tree. It has flaps of skin between its fore and hind limbs and uses its bushy tail as a stabilizer.

Daedalus and Icarus

stand that a human is not, and cannot be, like a bird.

Modern aircraft exploit certain characteristics of bird flight, but the effort to imitate it led early inventors into a blind alley. They sought to follow the lead of Daedalus, to fly by attaching wings to themselves and then moving them to generate power. A machine that achieves powered flight by flapping its wings is called an 'ornithopter' and, despite innumerable attempts to build one, it does not work.

The Daedalus technique inspired many more, at least in mythology. Some deities flew without the benefit of wings, others had rudimentary or vestigial wings — like those attached to the heels of Hermes — but some, such as angels, were represented as bearing large wings attached at the shoulders. It was soon realized that such wings could never lift a human being. Those of angels leave the arms free, but in any practical design the arms would have to supply support and power. In effect, the arms would have to be modified into wings, as the fore limbs of birds have been. From time to time a brave soul experimented with wings of this type, but the attempts invariably and inevitably ended in failure.

Leonardo da Vinci (1452-1519) is one of the most popular of all historical figures. He contributed to many branches of science and possessed what would be regarded today as a truly scientific outlook. He accepted nothing on the authority of others and trusted only in his own observations, which were acute, and experimentation. Aviation was not a major concern, but he loved to set himself problems and try to solve them, and he devoted considerable efforts to designing a flying machine. It was an ornithopter in which the aviator lay prone, strapped to a frame, and used a system of pulleys and levers to flap the wings. There is no doubt that a human could make the wings of such a machine flap — but it could not possibly fly.

The concept behind the ornithopter is deeply flawed because it takes account of only one aspect — and that inadequately — of the ways in which a bird is modified for flight. It mimics the vertical motion of the wings, but makes little or no allowance for the flexibility of real wings and the ease with which the bird alters the angle between wing and air flow — the 'angle of attack' — to vary the amount of lift they

to them so the largest — the primary flight feathers — were on the outside. Finally, he shaped the wings to give them a curvature like that of a bird's wing. He tested the apparatus and found it worked well. Then Daedalus fitted Icarus with the wings he had made for the boy and taught him to use them.

Before they made their bid for freedom, Daedalus gave his son a final, pre-flight briefing. He warned him not to fly too low, lest the mist clog the feathers, nor too high, lest the warmth of the Sun melt the wax. With that, they departed and all went well until Icarus grew over-confident, forgot the warning, and flew above his safe operational ceiling. The wax melted, the wings fell apart, and the boy fell to his death. Daedalus made good his escape, arrived safely in Sicily, and hung up his wings as an offering to Apollo.

No one can say how this Cretan legend began but it describes a perennial human dream. People have always longed to fly and, not surprisingly, they have based their ideas of flying on their observation of birds.

Today, we have almost achieved our ambition. Hang gliders attach wings to human bodies and when an engine is added, to make a microlight aircraft, something very like a winged human can take to the air — if less elegantly and a good deal more noisily than Daedalus and Icarus. The first step along the long path that has led us to the realization of the dream was to underproduce and to reduce air turbulence across them. Nor does it, or could it, achieve the compromises needed to generate lift and forward power — thrust — with just one pair of wings.

A bird possesses the necessary equipment and its ability to use it efficiently has involved much development of the brain, along different lines from those pursued in mammalian evolution. Its adaptations go much further, however. The weight of its body is reduced to a minimum. There is little fat. The large bones are hollow and filled with air. The jaws are small and lack teeth. Food is digested rapidly and the diet provides the maximum possible energy. This high metabolic rate is associated with a body temperature, depending on the species, between 107.6 °F and 113 °F (42-45 °C) — higher than that of a mammal — a very efficient respiratory system and a rapid heart beat — about 500 beats per minute in a sparrow and more than 1000 in a hummingbird. The energy produced is used to work very large and very powerful flight muscles — and a bird produces about 5.3 watts of energy per ounce (0.186 W/g) of muscle compared with about 4.7 watts of energy per ounce (0.165 W/g) for human muscles. Even so, a bird weighing more than about 33 lb (15 kg) probably needs all the power available to it in order to sustain straight

and level flight, so it is not very manoeuvrable and has difficulty taking off — it needs a long take-off run. This is the average weight of a mute swan (a wandering albatross weighs up to about 20 pounds, 9 kg). A human, weighing, say, 154 pounds (70 kg) would require wings nearly five times larger than those of a swan and the muscles to operate them — but the wings and muscles would add greatly to the weight, necessitating even larger wings. It has been calculated that an adult human equipped with wings like an angel would need chest muscles about 6.5 feet (2 m) deep to operate them.

Humans possess none of these necessary qualifications for flight, and the ornithopter offers no route to obtaining them. Its complicated mechanism simply adds to the weight that must be lifted.

The ornithopter was also based on another fallacious notion — that a bird obtains lift by pushing downwards against the resisting air. In effect, it assumes the bird pushes itself upwards. In fact, however, the bird is pulled upwards from above. The aerofoil shape of its wings — and, to a lesser but not inconsiderable extent of its body as well — cause it to be lifted by the region of low pressure above it (see page 182).

The clue to this was always there, even before Daniel Bernoulli supplied a clear explanation of it. Anyone could have

This is a modern microlight aircraft. Perhaps this is as close as most people will ever come to being able to fly like a bird or an insect. At the end of the flight, the aircraft can be folded up and taken away to be stored in a garage or some other comparatively small space.

observed the soaring and gliding flight of birds, in which they can gain height and cover long distances with very little beating of their wings. John Wilkins (1614-72), one of the founders of the Royal Society of London, commented on it and suggested that one day humans might achieve that type of flight, and around the end of the seventeenth century, several unsuccessful attempts were made to build a glider that would carry a human passenger safely. Had inventors persevered with these designs and abandoned the idea of the ornithopter, they might have had some modest success. They still would not have been able to build an aeroplane, of course, because they had no suitable source of power.

Hot-air balloons

At 1.54 pm on 21 November 1783, the aircraft designers were overtaken by events. After many experiments and with careful planning, two young Frenchmen, Joseph and Etienne Montgolfier, launched an airborne vehicle carrying two human passengers, Jean-François Pilâtre de Rozier and the marquis François Laurent d'Arlandes. They took off from the grounds of the Château de la Muette, on the western side of Paris, and flew for about 25 minutes, covering a distance of some 6 miles (9.6 km), then landed safely and without damaging the craft. They had used only a small part of their fuel — straw — and Etienne is said to have remarked that the machine could have remained airborne for much longer had the two pilots not lost their nerve.

The device was, of course, a hot-air balloon and its construction was possible partly because the Montgolfier family owned paper-making factories and had been making paper for several generations. They knew a great deal about the tensile strength and weight of paper, and their balloon was made partly from paper and partly from fabric. They had overcome the difficulty of lifting a large weight by following a route entirely different from that of the ornithopter designers. Their success gave 'lighter-than-air' flight an edge over 'heavier-than-air flight' that was not lost until the great airships fell from favour in the 1930s — and may not have been lost yet, for new designs for much improved airships are now being developed. Indeed, for a long time, the competition was not

between craft that were lighter or heavier than air, but between rival lifting gases — hot air and hydrogen.

When a balloon is described as being 'lighter than air' the principle underlying its operation sounds self-evident. That was not how it seemed two centuries ago. The composition of the air was only starting to be analysed and the prevailing scientific theory regarded heat as a substance, 'phlogiston'. The Montgolfiers did not know that when air is heated it expands and becomes less dense. They believed they were mixing air with phlogiston and the trick was to mix them in the correct proportions.

The use of hydrogen rather than heated air exploits the different physical properties of two gases, but when the first hydrogen balloon was invented, the difference was not apparent. Ordinary air, air mixed with phlogiston, and hydrogen were three quite distinct substances. The first hydrogen balloon was flown by Jacques-Alexandre-César Charles and M.N. Robert, in December, 1783.

Ballooning fired the public imagination. The first balloon crossing of the English Channel was made by Jean-Pierre Blanchard and John Jeffries on 7 January 1785 and, in 1804, Joseph Louis Gay-Lussac and Jean Biot used a balloon to explore the atmosphere, reaching an altitude of about 23,000 feet (7015 m).

The military potential of ballooning was also appreciated and, in 1789, the French army established a balloon corps. The first air raid took place in 1849, when an Austrian balloon bombed Venice. Balloons were used for observation by both armies in the American Civil War — where one of the observers fighting for the Union army was the Count Ferdinand von Zeppelin — and in 1870-71, when Paris was besieged during the Franco-Prussian war, a fleet of 70 balloons was used to deliver mail to the city and evacuate people from it.

Invention of the airship

It was immediately evident, however, that a balloon suffers from a major disadvantage: it cannot be steered. Attempts were made to persuade the craft to behave like ships but sails, oars and paddle wheels could not be made to work. The difficulty was not overcome until 1852 when, on 24 September, Henri Giffard travelled by air from the Hippodrome in Paris to a destination he announced in advance. Giffard had fitted a balloon with a very small steam engine, weighing 350 pounds (159 kg), generating 3 horsepower and driving a propeller 11 feet (3.4 m) in diameter at 110 revolutions per minute (rpm). Giffard had invented the airship. The concept was an immediate success and developments followed swiftly. Airships dominated the early years of aviation, offering scheduled flights and, in their later years, standards of accommodation and catering which rivalled those provided on ocean liners.

The operating principles on which airships are based are not quite so simple as they sound. The weight is lifted primarily by means of a 'lifting gas' — a gas that is lighter than air and whose lifting capability is equal to the weight of a given volume of air minus the weight of an equal volume of the gas. At sea level temperature and pressure, the weight of hydrogen is about 7 per cent of the weight of a similar volume of air and, therefore, replacing air with hydrogen provides a useful amount of lift. Hydrogen is highly flammable and, from 1922, the United States allowed only helium to be used. Helium is rather heavier than hydrogen, so it provides less lift, and it is more difficult and expensive to produce, but it was believed that its complete non-inflammability more than compensated for this disadvantage.

The weight the gas must lift comprises the sum of the weights of the airship itself, its engines and their fuel, the passengers or cargo and also the weight of such snow and moisture as may accumulate on the exterior of the ship during its flight. The weight does not remain constant. As fuel is consumed, the weight decreases and the weight may increase or decrease with changes in the temperature of the lifting gas. Extreme changes are compensated for by acquiring or releasing ballast or gas.

An airship is steered by large rudders and elevators, movable surfaces attached to the fins and stabilizers respectively. The 'cigar' shape facilitates control, and the hull and control surfaces also generate lift when the craft moves at an angle to the direction of the air flow.

The Goodyear airship, which has become a fairly common sight at major events, is, technically, a 'blimp' — the main envelope is not rigid and the shape of the craft is maintained by the pressure of gas. In a semi-rigid design, the envelope incorporates a rigid, longitudinal keel, and in a rigid design the hull consists of a framework covered by a fabric skin. Inside the hull, the gas is held in a series of cells, each of which can expand or contract, and those in the nose and tail are often separated from the main volume and used to 'trim' the attitude of the airship.

Inevitably, airships are large. The first Zeppelin, which flew in 1900, was 420 feet (128 m) long, 38 feet (11.6 m) in diameter, and was lifted by 388,140 cubic feet (10,990 cu m) of hydrogen. This was small compared to later airships. The largest was the *Hindenburg*, built to a Zeppelin design, which first flew in 1936. It was 804 feet (245 m) long and was lifted by 7.063 million cubic feet (200,000 cu m) of hydrogen. It was powered by four Mercedes-Benz engines, each developing 1000 horsepower and, at its cruising speed of 78 miles per hour (125.5 km/h), its range was 8750 miles (14,080 km). It carried 50 passengers. The *Hindenburg* flew between Germany and the United States, taking about 65 hours for the west-bound journey and 52 hours to cross from west to east — about half the journey time of an ocean liner.

The last flight of the *Hindenburg* ended, on 6 May 1937, with a disaster that claimed the lives of 36 passengers when the airship caught fire while it was docking at Lakehurst, New Jersey. The fire was spectacular but hydrogen is less dangerous than the *Hindenburg* conflagration made it appear — the victims died by falling, not burning. Because the gas is light, it rises as soon as it is freed from the envelope, so that the fire, burning very rapidly, moves upwards and away from the people below. The fire hastened the end of the airship era but did not cause it. Already, aircraft rather than airships were flying regular air services within Europe and the United States, and between the United States and South America as far south as Argentina. A transatlantic service was opened in 1939 and Imperial Airways was founded on 1 April 1924, to provide a link between Britain and the Middle East. The route was extended to Karachi in 1929 and to Brisbane in 1935.

Aeroplane development
Aeroplanes carried somewhat fewer passengers than airships in much less comfort, but

Hindenburg, the largest airship ever built.

they were fast. The world air speed record was set in 1931 by the Supermarine S 6B seaplane at 407.5 miles per hour (656 km/h), but fare-paying passengers might travel in the all-metal, twin-engined DC-3, or Dakota. Arguably the safest and most reliable aircraft ever built, it entered commercial service in 1936 — and a few are still flying. It carries up to 21 passengers at a cruising speed of 180 miles per hour (290 km/h).

The victory of the aeroplane over the airship was not won easily because, for a long time, the time saved by travelling faster was lost by the short range of aeroplanes and the consequent need for frequent refuelling. Today, negotiating permissions for a new long-distance air route consists mainly in clearing the route with the relevant air traffic control centres and international regulatory authorities. In the early days it was a matter of providing a series of landing sites with refuelling facilities.

The speed of an aeroplane is made possible by its streamlining but, in a sense, the practical and commercial advantages it brings are a bonus. An aeroplane must fly fast or it cannot fly at all. The difference in air pressure over the upper and lower surfaces of an aerofoil is produced by the flow of air. The greater the flow of air, the greater the difference in the two pressures and, therefore, the more lift that is produced. To develop enough lift to raise its weight, an aeroplane must move fast enough to generate the necessary air flow over its lifting surfaces. That is why a powered aeroplane needs a take-off run, and a glider must either be towed to gain flying speed or must be launched from a great enough height for it to accelerate to that speed as it falls.

In no sense did the aeroplane evolve from the balloon or airship. It followed an entirely different path that began very soon after the flight of the Montgolfier balloon.

Once designers had finally abandoned all hope of making a working ornithopter, they were free to concentrate on the possibilities of soaring and gliding flight based on wings that were fixed rather than flapping. This led, in turn, to the development of the aerofoil wing.

As long ago as 1799, Sir George Cayley (1773-1857) had designed what would eventually become the modern aeroplane. It had a fuselage, fixed wings, and a tail unit that combined elevators and a rudder and he recognized, and solved, the problems of maintaining stability, partly by using dihedral — wings that slope upwards so the tips are higher than the roots (the opposite is called 'anhedral'). He designed cambered wings — aerofoils — to enhance lift, considered the merits of biplane and triplane designs — which were built in 1868, after Cayley's death. He realized such a craft would need an undercarriage for landing, and the lightweight wheels he designed for the purpose later provided the basis for the spoked bicycle wheel. He also knew that an aeroplane — the word 'aeroplane', literally a flat surface (plane) in the air, is believed to have been coined in 1866 — would require an engine, so he designed one of those, as well. In 1853 a glider built to his design, with the controls locked, made the first flight carrying a human passenger — Cayley's coachman.

Many other inventors were in pursuit of the ancient dream — to fly like a bird. Perhaps the most successful of them was Otto Lilienthal (1848-96), a German engineer. He made hundreds of flights in gliders he had designed until he died when one of them crashed.

All that remained was to find a suitable means of propulsion. The screw propeller, already in use in ships and being introduced in airships, was generally accepted as the best means of obtaining forward motion — thrust. The propeller accelerates the air passing it, producing a force acting towards the rear. Because any force applied in one direction must be balanced by an equal force in the opposite direction — Newton's third law of motion — the rearwards force acting on the air is balanced by a forward-acting force on the propeller itself and, therefore, on the whole aeroplane. The problem lay not in the theory of propellers, but in finding an engine powerful enough to deliver the necessary thrust yet light enough to be carried. Airship engines were available, and included the first airborne internal combustion engine, which flew in 1872, but they were inadequate for the task. An aeroplane must use power to provide its lift

A sailplane, designed for maximum aerodynamic efficiency. Its long, narrow wings produce a great deal of lift at low speeds.

The space shuttle about to be launched. Its rocket engines carry it beyond the regions where thrust can be produced by compressing the air. Its rudimentary wings and generally aerodynamic design allow it to glide back to the surface.

as well as its propulsion. The first controlled flight of a powered aeroplane, made by Orville and Wilbur Wright on 17 December 1903, used a four-cylinder gasoline engine designed and built by the brothers themselves. It weighed 179 pounds (81.3 kg) and developed 12 horsepower.

The necessary components of airframe, controls and engine had come together, and the subsequent history of propeller-driven aircraft consists of a steady progression of refinements, improvements and adaptations to particular requirements. Very early in this story, however, some designers diverged from the main stream to pursue a different idea.

Lift is produced by passing air over an aerofoil and conventionally it was achieved by moving the entire aeroplane, to which the aerofoil is fixed. If the wings were able

to rotate independently of the airframe, however, they would produce lift in an aeroplane that was not moving horizontally. The machine — a helicopter — could rise and descend vertically. The Chinese had played with toys of this kind for centuries; Leonardo da Vinci and Sir George Cayley were just two of many people who considered this type of design; and, in 1909, Igor Sikorsky (1889-1972) built one that lifted its own weight but had insufficient power to carry passengers or a payload. Several German helicopters flew in the 1930s and, in 1940, Sikorsky produced the prototype of the design that stimulated most of the helicopter developments which followed.

A propeller blade travels much faster than the aeroplane to which it is attached and this sets a limit of around 400 miles per hour (644 km/h) to the speed of propeller-driven aircraft. The jet engine, invented independently by Hans von Ohain in Germany and Frank Whittle in England, overcame this difficulty. Instead of pushing air to the rear, air is compressed and then heated inside combustion chambers, causing it to expand, and it is the expanding air that provides the rearward acceleration and, therefore, the forward thrust. The first jet-powered aircraft flew in 1939 in Germany and 1941 in England.

Conventional engines, of both types, use air to burn their fuel and to produce their thrust, which confines them to the denser regions of the atmosphere. The rocket engine escapes this limitation by generating thrust through the expansion not of air but of its own exhaust gases, and the vehicle powered by rocket engines can carry its own supply of oxidant to burn the fuel. Independent of the atmosphere, increasingly powerful and sophisticated rocket engines made possible the exploration of space.

On 23 August 1977, in California, designer Paul MacCready and pilot Bryan Allen realized the final dream, of human-powered flight, in *Gossamer Condor*, powered by a propeller driven by pedals and a bicycle chain. The same team flew 23 miles (37 km) across the English Channel on 12 June 1979, and on 23 April 1988, Kanellos Kanellopoulos flew 76 miles (123 km) from Heraklion, Crete, to Thera in 3 hours 54 minutes. The first pedal-powered helicopter, flown by Greg McNeil, hovered 8 inches (20 cm) above the ground for 6.8 seconds on 12 December 1989, at San Luis Obispo, California.

Some of the earliest pictures of flying machines depict airborne ships — how they remain airborne is uncertain — that are being driven by sails. It is doubtful whether anyone ever attempted to build a vessel powered in this way, because it is obvious that it could not work. A balloon moves with the air mass that contains it and, therefore, sails could make no contribution. A vessel on the surface, however, can be propelled by sails, because, the vessel being in contact with a surface that is stationary, they exploit the motion of the air in relation to the surface.

Wind for propulsion

Several unrelated groups of animals are able to glide or fly and, despite the physiological modifications it demands, the evolution of

flight may well be inevitable on a planet with a suitable atmosphere. The exploitation of the wind as a means of propulsion is a very different matter. It offers no advantage to land animals. They use legs for locomotion and, when an animal runs at full speed, it is already moving its legs as fast as they can move. Pushing it to move faster would not help.

Aquatic animals, however, are in a different position and several unrelated surface-dwellers have evolved sails. *Vellela*, for example, the 'by-the-wind-sailor', is not a single animal but a colony of specialized hydroids, one of which is greatly enlarged and shaped like a sail which carries the colony before the wind. The Portuguese man-of-war (*Physalia physalis*), also a colony rather than a single individual, sails in the same way. The sailfish (*Istiophorus platypterus*) has an extremely large dorsal fin. When the fish is swimming — it is a fast, powerful swimmer — the fin is folded into its back, but at other times the fish travels more slowly, its fin extended above the surface and functioning as a sail. Several species of whales economize on energy by raising their tail flukes — which are horizontal and not vertical like the tail fins of a fish — above the surface and using them as sails for slow, relaxed cruising.

When our ancestors invented sails, they had no need to study marine animals. A person walking on a windy day can feel the pressure of the wind, especially when wearing voluminous garments. In any case, no one could observe marine animals before sea-going ships — driven by sails — came into use. Sails were certainly used on boats plying the Nile, perhaps as long ago as 4000 BC and, by about 1500 BC, sails were being used, often to augment rowers, on ships that were widely engaged in trade and warfare in the eastern Mediterranean. According to Greek legend, the departure for Troy of the army and fleet assembled by Agamemnon at Aulis, in Boetia, was delayed by the goddess Diana who, offended because Agamemnon had killed a stag sacred to her, produced a calm that prevented the ships from leaving port. Only when the King offered to sacrifice his daughter, Iphigenia, did the goddess relent. Archaeological evidence suggests that Troy may have been destroyed by invaders some time around 1200 BC, by which time the Greeks were conversant with the use of

wind propulsion and, it seems, no less familiar with its unreliability.

The development of sailing ships has been much concerned with overcoming this unreliability. Increasing the total area of sail, by equipping vessels with bigger sails and more of them, made it possible to exploit lighter winds, but this was of little value until a means had been found for sailing in a direction other than that of the wind itself. On Roman merchant ships, which carried no rowers, the sails were controlled by shrouds — a net-like system of ropes — by which they could be swung around the mast, allowing the ship to sail at an angle to the wind and, by the twelfth century, shrouds were being used in northern Europe. The major advance came around the end of that century, with the replacement of rudders developed from steering oars by the rudder that is fixed to a stern post. This improved greatly the efficiency with which ships could be steered against the direction of the wind.

The design of sailing ships reached its peak by the middle of the nineteenth century, with the clippers. Their sleek hulls, up to five times longer than they were wide, and extremely large sail area made them very fast, and it was not unusual for a clipper to cover more than 400 miles (644 km) in a single day. In the end, of course, sailing ships could not compete economically against steam ships, but they were not replaced rapidly. They needed no storage space for fuel and required no refuelling stops, so what they lost through their slower speed and the vagaries of winds they gained in lower operating costs, fewer stops and faster turn-around times. A few Finnish-owned sailing ships were still in commercial use in the late 1940s.

From time to time there has been talk of re-introducing sailing ships in response to rising fuel prices, and a number have been designed. That they have so far failed to enter commercial service on a large scale is probably due mainly to the fact that the rises in transportation costs, which might render them economically viable, are invariably accompanied by a reduction in seaborne trade and, therefore, a surplus of ships competing for cargoes. Nevertheless, a few wind-assisted, diesel-engined ships have appeared, such as the *Patricia A*, for example, designed by the Dynaship Corporation, of Palo Alto, California, and working in the Caribbean, and a Japanese oil tanker, the *Shin Aitoku Maru.*

The new ships take advantage of electronic equipment that has become standard since sailing ships were last in commercial use. In the past, for example, the master of a ship had to interpret the conditions of the sky and sea, relying on his own knowledge and experience to predict the location of favourable winds, and then plan a route accordingly. A modern ship has constant access to information from coastal stations, weather ships and meteorological satellites and computer programs to calculate routinely updated headings and feed them directly to the steering mechanism. The position of the ship, determined by reference to navigational satellites, is displayed on a screen and remote sensing equipment, such as radar, removes the hazards once associated with poor visibility.

The modern sailing ships are powered by the wind, but their appearance departs radically from that of their forebears. The *Patricia A*, 170 feet (52 m) long and displacing 450 tons, carries four sails which are set, in winds of less then 40 knots (74 km/h), or furled at higher wind speeds, by means of powered rollers inside hollow masts and the masts themselves rotate to maintain the optimum angle between the sails and the wind. The masts and sails are computer controlled from the bridge, engine and sails being deployed to maintain a constant speed with the minimum use of diesel fuel. The *Shin Aitoku Maru* carries two rigid, plastic sails, shaped like upright half cylinders. Each sail is made in three panels and the outer panels can be folded back around the mast to 'furl' the sail or to alter the area exposed to the wind.

In other designs, sails made from fabric have been replaced by two alternatives. In one, the 'sails' are of fairly conventional shape but consist of rigid, horizontal struts that can be rotated on their own axes, like the struts of a venetian blind, while the whole structure can be rotated about the mast. The setting and furling of sails is replaced by opening or closing the 'blinds'.

An even more radical design exploits the fact that when a smooth-surfaced cylinder rotates in flowing air, at about four times the speed of the air flow, the air pressure is greater on one side of the cylinder than on the other. The difference can be increased about tenfold if discs of a larger diameter

(Opposite) The days of sail are gone, but sailing remains a popular sport. Modern fabrics and designs allow racing yachts to exploit the wind much more efficiently than was possible in the past.

than the cylinder are fixed at the top and bottom. This is the 'Magnus effect' and it generates a force on the cylinder acting at 90° to the direction of the air flow. Instead of sails, a 'Magnus' ship carries tall, rotating cylinders and uses its rudders and variations in the rotational speeds of its cylinders to maintain its heading, virtually regardless of the wind direction. The principle underlying such a ship was discovered in 1922 by a German engineer, Anton Flettner, at the University of Göttingen. He built a small rotor-powered boat and conducted larger sea tests on the *Buckau*, a 960-ton schooner, replacing two of the masts by 42-foot (13-m) masts around which the cylinders rotated. The cylinders were turned, at 125 revolutions per minute, by small motors and varying the speed of one or other rotor and, therefore, the thrust it delivered, caused the ship to turn. In 1925, Flettner crossed the Atlantic in his rotor-driven ship.

Sails that were hung from yard arms and moved by ropes and pulleys required ships to carry large crews, often working in very dangerous conditions. Steamships could be operated by smaller crews and today, when most of the necessary operations are automated and controlled by computers, even the largest cargo vessels can be managed by a handful of sailors. Computers also enable new sailing ships to retain this advantage.

Wind power

Where there is a large expanse of smooth, uncrowded sand you may see enthusiasts racing sand yachts at respectable speeds. These are the only land vehicles to harness the propulsive power of the wind. It is not well suited for use on roads, because headings are predetermined and cannot be adapted to prevailing conditions. The wind has driven machinery for centuries, however, and is seen increasingly as a means for generating electrical power.

The first windmills, based on principles developed for watermills, were in use in the seventh century in what is now Iran. It is possible that Hammurabi (about 1792-50 BC), the Babylonian emperor, planned to use them much earlier for pumping irrigation water, but there is no record of the mills having been built. Early in the thirteenth century, the armies of Genghis Khan took captured millwrights eastwards to instruct local people in the necessary skills, and soon after that they were to be found throughout

Europe and Asia.

For most of their history, windmills have been used either to pump water or to grind corn. Many of those for which the Dutch landscape is famous pump water to drain land that has been reclaimed from the sea and lies at or below sea level. The first mill to generate electrical power was built in 1890, in Denmark, in what was then the most advanced form of the traditional style. Although modern wind generators are substantially different in appearance, and

much more efficient, their underlying principles remain the same.

In a windmill or wind generator, the wind causes the rotation of 'sails' — in fact, rotor blades — and the rotary motion drives the machinery. The axis around which the 'sails' turn may be vertical or horizontal. The earliest windmills had vertical axes, but over the centuries, and despite its greater complexity, the horizontal-axis type has proved more efficient and it has become the traditional windmill.

In a vertical-axis mill, the 'sails' are held on radial arms and, in the simplest version, the axis is fixed directly to a grindstone. A horizontal axis, on the other hand, requires gearing to harness its rotation, because the size of the 'sails' means the axis is high above the ground. The 'sails' will turn only when they are at right angles to the wind and, therefore, an additional mechanism is required to turn the sails and their axis as required.

The earliest mills achieved this by

The *Clifton Flasher*, a racing catamaran, is equipped with modern, rigid sails.

The largest wind
generator in Britain, in
Orkney, stands 246 feet
(75 m) tall and gener-
ates 3 megawatts of
power.

housing the entire mechanism, including the corn-grinding machinery, inside a building that was free to rotate about a strong post. The building stood on top of a lower, fixed building, the base of which was usually but not always at ground level, and the post passed vertically through both. This is a 'post mill'. 'Tower mills', which began to be built early in the fifteenth century, housed the horizontal axis in a cap that could be turned on the top of what was otherwise a tall, solid building — a tower — and the rotating axis was linked by gearing to the machinery below.

The upper part of both types of mill was turned by means of a long 'tailpole'. At first the tailpole was hauled by hand and later by winches until, in 1745, Edmund Lee invented the 'fantail'. This is a set of vanes mounted on the tailpost at right angles to the 'sails' and linked by gearing to wheels on the base of the tailpost that run in a circular track around the mill. When the wind shifts direction, so the 'sails' are no longer at right angles to it, the vanes begin to turn. This drives the wheels and moves the mill around until the vanes are no longer exposed to the wind — but the 'sails' are.

Most modern wind generators are direct descendants of the tower mill, although vertical-axis designs have also been developed. The 'tower mill' generators consist of rotor blades, which are no longer called 'sails', mounted on horizontal axes and set on the top of towers high enough for the blades to be clear of the surface boundary layer, the region near the ground where the wind is slowed by friction and made turbulent by obstructions. In small versions the rotors are held into the wind by a rudder at the end of the tailpost, those of large generators by small motors controlled by wind sensors and computers. The idea of 'harnessing the wind' has obvious attractions and, over the last century, it has engaged the attention of many inventors and, because the technology seems straightforward and can be applied most easily on a small scale, these have included many amateurs as well as professionals.

Despite its ancestry, a wind generator performs under very different economic circumstances and its task is different, and mechanically much more difficult, than that of a windmill. The windmill had no real economic competition. It did not compete with its ancestor, the watermill, but pro-

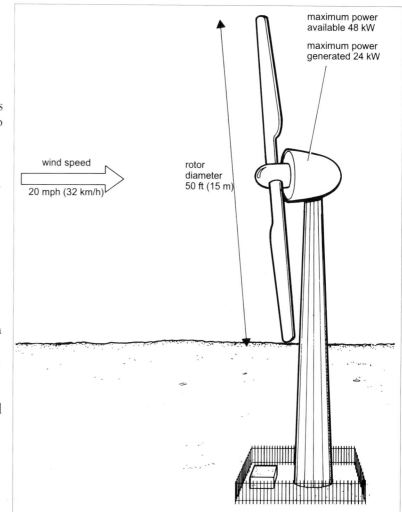

wind speed
20 mph (32 km/h)

rotor diameter 50 ft (15 m)

maximum power available 48 kW

maximum power generated 24 kW

vided an alternative where water power was not available, taking over tasks that had previously depended on the muscle power of draught animals, liberating them for more productive use in the fields. The wind generator, on the other hand, competes against a firmly established industry with many years of experience that operates well below the capacity of its existing plant.

The grinding of corn and pumping of water require slow-moving machinery. An electric generator, on the other hand, must turn at a high speed — measured in thousands of revolutions per minute — and its output is directly related to its rotational speed. A wind generator, therefore, must incorporate gearing to increase greatly the speed its rotors deliver. This raises a problem, because a significant proportion of the energy delivered by the rotors must be expended in driving the gearing itself. The loss is inevitable and to compensate for it, the power from the rotors must be maximized. This can be achieved partly by

The power produced by a wind generator is proportional to the diameter of the rotor and the speed of the wind. In a 20-mph (32-km/h) wind the power available to an aerodynamically efficient 50-foot (15-m) rotor is about 48 kilowatts (kW), but the electrical power generated is only half that amount, because of losses in the gearing and generating mechanisms.

making the rotor blades aerodynamically efficient, and the history of aircraft design has provided ample experience of the behaviour, under different airflow conditions, of aerofoils in general and propellers in particular. The rotor blades of a modern wind generator are very advanced, specialized, propellers.

The amount of energy a propeller can extract from the wind is proportional to the square of the diameter of the circle it describes and the cube of the wind speed. This has two consequences. It means that generators must be sited in windy locations and they must be high above the ground, because wind speed increases with height. It also means the rotors must be very large, to sweep the greatest possible area. Wind generators, therefore, are tall. Most stand on towers about 80 feet (24 m) high and some are twice that size or more. The United States National Aeronautics and Space Administration (NASA) has designed one with a rotor diameter of 200 feet (61 m).

Designs for rotors based on aircraft propellers began to appear in the 1930s and, in 1944, the Germans planned to build a series of wind generators with a rotor diameter of 400 feet (122 m). These were never constructed but, in 1940, one giant was, on a hill in Vermont called Grandpa's Knob. It had metal blades with a diameter of 87 feet (26.5 m) and stood on a tower 117 feet (36 m) tall. It ran for several years before it was destroyed by metal fatigue in one of its blades.

Nowadays, large generators are equipped with rotors made not from metal but from advanced materials that are stronger, lighter and less prone to fatigue. Even so, formidable engineering problems have had to be addressed to build structures of this size that will turn in a very light wind and brake automatically when the wind speed approaches their safety threshold.

Vertical-axis designs are derived mainly from the work of G. Darrieus, in France, and S.J. Savonius, in Finland. The Darrieus design uses a variable number of aerofoils that are mounted vertically. These may be bowed, so they sweep outwards at the centre and are attached to the axis at top and bottom, or straight and held at the end of arms with a mechanism to alter the angle between the blade and the arm. The Savonius design consists of two semi-cylindrical sheets mounted with their

concave sides facing one another, one on either side of the rotational axis. The principle is the same as that of the panemone — the curved plate that spins in the wind outside shops and filling stations to attract customers and advertise products. A Savonius rotor is sometimes incorporated into a Darrieus design to help start the large rotors turning at low wind speeds.

In the early 1930s, a Brazilian inventor, J. Madaras, planned a wind generator based on Flettner rotors. It was never built, but would have comprised 40 cylinders 90 feet (27.5 m) high and 28 feet (8.5 m) in

diameter that would have moved around a circular track 1100 yards (1 km) in diameter. Electric motors would have spun the rotors, their direction of spin reversing twice during each circuit, the wind would have driven the rotors around the circuit, and power would have been generated from the wheels on which the rotors moved.

While most designers have sought to build ever larger rotors, others have tried to devise ways of making the wind blow faster. This is possible, because — as anyone who has walked between high buildings knows — wind can be funnelled. One such design,

by Alfred Weisbrich, employs a stack of hollow, toroidal (doughnut-shaped) rotors in which the outer edge is open and flanges are positioned inside each torus. The wind is accelerated inside the torus. Another, by Pas Sforza, uses a triangular aerofoil mounted at an angle to the horizontal with small rotors on top of the triangle, at its base. The shape of the aerofoil is said to increase the wind speed by 50 per cent or more. A funnel-shaped shroud fixed behind a rotor will also accelerate the wind. As the air passes behind the rotor it expands inside the funnel. This reduces the air pressure,

A wind farm in California. It has been estimated that to generate 10 per cent of its electricity Britain would need 30,000 to 40,000 generators like these, occupying about 1250 square miles (3238 sq km) of land.

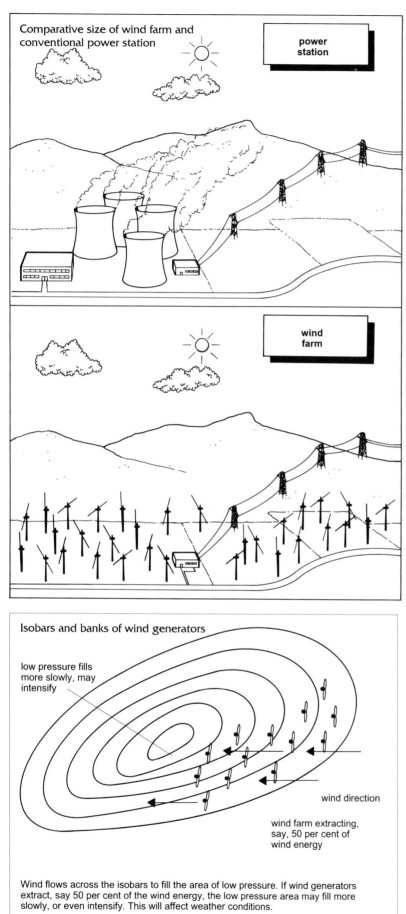

Comparative size of wind farm and conventional power station

power station

wind farm

Isobars and banks of wind generators

low pressure fills more slowly, may intensify

wind direction

wind farm extracting, say, 50 per cent of wind energy

Wind flows across the isobars to fill the area of low pressure. If wind generators extract, say 50 per cent of the wind energy, the low pressure area may fill more slowly, or even intensify. This will affect weather conditions.

drawing more air past the blades.

Some designs are much more ambitious. One, for example, would mount rotors like aeroplane propellers on the wings of gliders, which would then look much like multi-engined aircraft. The gliders would be flown tethered, like kites, stacked one above the other, their aircraft shape ensuring they always faced into the wind. What is perhaps the biggest concept of all, however, would accelerate the wind by creating an artificial hurricane. A very large turbine would be housed at the apex of a cone set at the base of a wide, cylindrical tower, nearly 2000 feet (610 m) high. The top of the tower would be open and in its sides there would be shutters that could be opened to face the wind, allowing air to enter. Inside, the shape of the tower, would force the wind to spiral upwards, creating a vortex with a region of very low pressure at its centre. This would draw in air from below the tower, where it would be funnelled past the turbine.

Such devices aim to overcome the two greatest obstacles to the harnessing of wind power. Although the movement of air represents a vast amount of energy, the energy is very diffuse and the movement is not constant. Most wind generators intended for domestic use produce no more than about 3 kilowatts (kW). One bar of an ordinary radiant electric fire uses 1 kW, so this is not very much. The very largest generators, feeding the public supply, produce up to about 1.5 million watts (MW). In both cases, the actual output depends on the wind speed. Existing British power stations have a total 'installed capacity' — the production of which they are capable — of almost 65 thousand million watts (gigawatts, GW) and those of the United States of 743 GW.

The generation of electricity is not difficult but electricity cannot be stored. It can be converted to a different form of energy and converted back into electricity when required — as in a battery, where it is converted to and from chemical energy — but energy is wasted in the conversion. Conventional power stations generate electricity for immediate use and, so far as possible, wind generators must do the same. They can do so in useful amounts only if large numbers of them are installed in arrays — as 'wind farms'.

A typical modern electricity generating station has an installed capacity of about 1 to 1.5 GW (thousand million watts) and each year the industry generates a total of about 300 GW in Britain and about 2700 GW in the United States. This is about half the British and 40 per cent of the American capacity. Less than half of the electricity generated is used in private homes. In most industrial countries, including Britain and the United States, about 35 per cent of the electricity generated is used domestically and about 65 per cent is for non-residential use and, in some countries, the difference is even more marked. In Turkey, for example, only 14 per cent of electricity is used residentially and 85 per cent non-residentially, and in Japan the figures are 20 per cent and 80 per cent respectively.

Wind farms are large because of the nature of the market they seek to enter. They are regarded as alternatives to conventional power stations and, therefore, they must achieve similar levels of output. The output of a single power station is approximately 1000 times more than that of a large wind generator — the Grandpa's Knob wind generator had a 1.25 MW capacity. While it is practicable for one or two wind generators to serve an isolated, community — such as the population of an offshore island — that lies beyond the economic reach of the grid supply network, it requires 1000 to 1500 of the large wind generators — and twice as many of the smaller, 500-750 kW, types — to equal the capacity of a large power station and it is misleading to suggest that a proposed wind farm will supply a particular number of homes, unless reference is also made to the much larger supply required for non-residential use, especially if there are local manufacturing industries.

The rotors of a wind generator can turn to 'find' the wind but, in a wind farm, the position of each individual generator must be calculated with careful regard for the direction of the prevailing wind. This is because a wind generator extracts energy from the wind and, therefore, the wind is slowed by it and the generator screens any other generators situated behind it, reducing their efficiency. Provided no other generators interfere, the wind speed will recover by mixing with 'untapped' wind after a distance approximately equal to eight times the rotor diameter of the generators, but the spacing can be reduced by staggering rows of generators to minimize their mutual interference. There must also be adequate horizontal

spacing — of about 400 yards (366 m) between each generator. The wind farm, therefore, consists of staggered rows of quite widely spaced generators, each occupying about 5 acres (2 ha), and a farm with the capacity of a conventional power station is likely to have a total area of 5000 to 10,000 acres (2020-4050 ha). Some engineers have suggested that the generators should be much more widely spaced to compensate for very local variations in wind, and alternative schemes involve siting a small group of generators on the top of each of a range of hills with all of the hilltop groups linked.

Between the generators, and at a lower level, pylons connect the output to the grid network. For reasons of safety and security, humans must be barred from unauthorized entry to the land occupied by wind farms, although some of the land can be used for livestock grazing.

Few sites offer such a large, open, windy area — although there are suitable ranges of hills — and, consequently, few wind farms

As offshore oil fields are depleted, wind generators might be installed on redundant oil-production platforms to exploit the generally windy conditions that prevail over the sea.

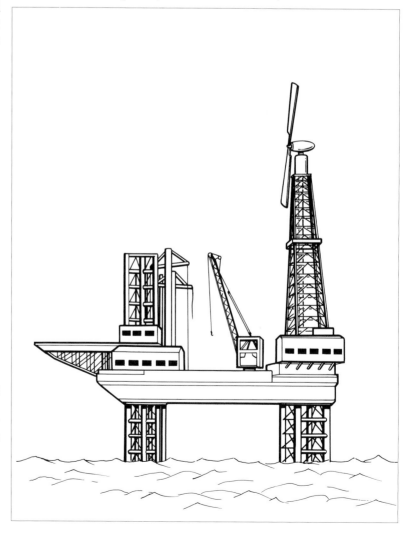

have been built. It might be feasible, but expensive, to construct offshore wind farms, perhaps utilizing and linking oil production platforms when their oilfields are exhausted. The few farms that have been built so far are generally small and their contribution to power supplies is minor.

Despite their large land requirements, their effect on the aesthetic quality of landscapes and the large manufacturing industry, using substantial amounts of physical resources, that would be needed to mass produce the generators, wind farms are held to be environmentally harmless. In particular, they are supposed to have no influence on climate. While wind farms remain small and few in number, this is probably true. On a large scale, however, it might not be.

The weather we experience at the surface results from the interactions among the conditions which produce pressure differences, the characteristics of the resulting air and the restoration of equilibrium as depressions are filled by air that spills from regions of higher pressure. Wind is the movement of air around areas of atmospheric high and low pressure and its effect is to equalize pressure. Usually, this happens in a matter of a few days or a week.

The purpose of a wind generator is to extract as much as possible of the energy of motion — kinetic energy — of the wind. The efficiency of the rotor determines the proportion of that energy which can be captured. This varies according to the wind speed, but in a wind of about 20 miles per hour (32 km/h) it is around 40 per cent (this is not the overall efficiency of the generator, because of losses in generation and transmission). It is reasonable to suppose that an array of generators might increase this efficiency to at least 50 per cent by exposing several rotors to the same wind.

The consequences of wind power

It is possible, therefore, that a large wind farm, comprising more than 1000 very large rotors strategically deployed to maximize the efficiency of the total array, might reduce the wind energy by half over a considerable area. Weather systems move, of course, but each system leaving the wind farm would be significantly depleted. This would delay the equalization of high and low pressures. Clearly, the weather would be affected locally. Most obviously, people living 'downstream' of the wind farm might experience generally gentler winds but beyond that it is difficult to predict what the consequences might be. If the equalization of pressures is delayed, but the conditions producing the differences remain, perhaps the pressure systems would intensify. People might experience prolonged episodes of weather of a particular type that ended with a brief spell of very violent weather, heralding the establishment of another episode of a different type.

Very large arrays of passive solar or photovoltaic collectors might also produce meteorological effects. These devices work by absorbing solar radiation and, therefore, they have a very low albedo. The devices are designed to utilize as much as possible of the energy they absorb and, therefore, they will not emit a significant amount of long-wave, black body radiation. If they are sited in a region of high albedo, however, such as a sandy desert or a snow-covered surface, and if a particular area contains a large number of arrays, the overall albedo might be reduced throughout the area. This would alter the extent of warming or cooling of the air in contact with the ground and, therefore, the weather 'downstream' of the array.

In our anxiety to minimize the adverse environmental consequences of our activities we should beware of supposing there are alternatives which have no environmental effects at all. Everything we do will have some environmental effect, and the choices we make must be determined by the most careful, informed calculation of the relative advantages and disadvantages.

Solar cells and albedo. A large array of solar collectors may alter significantly the local albedo, especially if it is sited in an area of high albedo, such as a desert. Altering the albedo will create a 'hot spot', possibly affecting local weather.

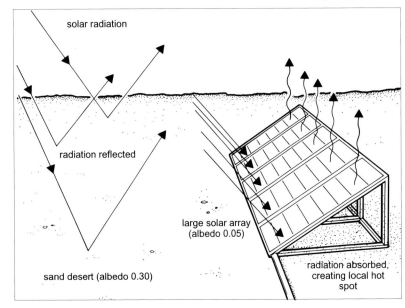

solar radiation

radiation reflected

large solar array (albedo 0.05)

sand desert (albedo 0.30)

radiation absorbed, creating local hot spot

INDEX

Acknowledgements

The publishers wish to thank the following for supplying photographs for this book:

Page 2 NASA; 3 NASA; 6 Mary Evans Picture Library; 8 Mary Evans Picture Library; 9 Mary Evans Picture Library; 11 Hutchison Library/ Pierrette Collomb; 15 Hutchison Library/J.V. Puttkamer; 16 Science Photo Library (SPL)/Pearson/Milon; 17 Dr. A.C. Waltham/Nottingham Polytechnic; 18 Hutchison Library; 19 Oxford Scientific Films (OSF)/Anna Walsh (top), RIDA/David Bayliss (bottom); 20 David Attenborough (top), Hutchison Library/P. Goycolea (bottom); 21 Hutchison Library/John Egan; 22 Premaphotos Wildlife/M. Preston-Mafham; 24 Hutchison Library/P. Goycolea; 25 Hutchison Library; 29 Hutchison Library/Jeremy Horner; 31 OSF/Martyn Chillmaid; 32 Planet Earth Pictures/P.J. Palmer (top left), Planet Earth Pictures/Geoff du Feu (top right), Planet Earth Pictures/Mark Mattock (middle left), Planet Earth Pictures/P.J. Palmer (middle right), Planet Earth Pictures/Ken Lucas (bottom); 34 OSF/Breck P. Kent; 35 OSF/Peter Parks; 36 RIDA/David Bayliss; 37 Hutchison Library/Hilly Janes; 38 OSF/W.J. Kennedy; 39 Dr. A.C. Waltham/Nottingham Polytechnic; 43 Premaphotos Wildlife/R.A. Preston-Mafham; 46 NASA; 49 SPL/ NASA; 52 RIDA; 53 SPL/US Geological Survey (top), SPL/NASA (bottom); 54 Chris Bonnington/Doug Scott (top), SPL/Michael Gilbert (bottom); 55 Hutchison Library; 58 SPL/Sabine Weiss; 61 Robert Harding Picture Library; 62 The Royal Aeronautical Society; 64 The Royal Aeronautical Society; 66 Coulsdon Library; 67 The Royal Aeronautical Society (top), The Royal Aeronautical Society (bottom); 69 Jodrell Bank; 71 Hutchison Library/Julia Davey; 72 Hutchison Library/Brian Moser; 76 Dundee; 81 RIDA/R. Sandman; 88 D.M.G. Buchanan (top), C.S. Broomfield (bottom); 89 C.S. Broomfield (top), J.F.P. Galvin (bottom); 94 OSF/Kim Westerkov; 100 OSF/Ben Osborne; 104 Hutchison Library/Timothy Biddow; 105 Hutchison Library/David Brinicombe (top), Hutchison Library (bottom); 106 Hutchison Library (top), Hutchison Library (bottom); 111 The Telegraph Colour Library/ESA; 113 G.V. Mackie; 116 University of East Anglia; 118 OSF/Jack Dermio; 119 OSF/Scott Camazine; 121 British Coal; 123 OSF/J.A.L. Cooke (top), Planet Earth Pictures/Storm Stanley (bottom); 124 Premaphotos Wildlife/K.G. Preston-Mafham (top), Ian Findlay (bottom); 129 Ian Findlay; 130 Hutchison Library/Julia Davey; 132 Institute of Archaeology (top), OSF/Richard Kirby (bottom); 134 Commissariat à L'Énergie Atomique; 139 SPL; 140 SPL/ National Centre for Atmospheric Research (USA); 150 Hutchison Library/Tim Motion; 152 Shell; 153 Environmental Picture Library/V. Miles; 154 Hutchison Library/Liba Taylor (top), Environmental Picture Library/Mike Jackson (bottom); 155 Environmental Picture Library/V. Miles; 157 OSF/Stephen Dalton (top), SPL/John Sanford (bottom); 158 OSF/E.R. Degginger (top), SPL/Phil Jude (bottom); 160 Robert Harding Picture Library; 161 NASA; 162 Hutchison Library/J.G. Fuller (top), Hutchison Library/J. Puttkamer (bottom); 168 Hutchison Library/Bernard Gerard; 171 Environmental Picture Library/Martin Bond; 173 Richard Garratt Design; 175 SPL/NASA (top), Premaphotos Wildlife/K.G. Preston-Mafham (bottom); 176 Hutchison Library/Nigel Smith (top), University of East Anglia (bottom); 178 Premaphotos Wildlife/K.G. Preston-Mafham; 181 OSF/Michael Leach; 182 OSF/Stephen Dalton; 183 OSF/Marty Stouffer Prods.; 185 Richard Garratt Design; 188 The Royal Aeronautical Society; 189 British Gliding Association; 190 The Royal Aeronautical Society; 192 Alastair Black; 195 Alastair Black; 196 Hutchison Library/Pierrette Collomb; 199 Hutchison Library/J.G. Fuller.